Algebra Experiments II
Exploring Nonlinear
Functions

Ronald J. Carlson
Mary Jean Winter

DALE SEYMOUR PUBLICATIONS
Pearson Learning Group

We would like to thank the many teachers whose enthusiasm for the algebra experiments has encouraged us. Special thanks to the participants in the Exeter Conferences whose suggestions we gratefully acknowledge. Thank you also to our colleagues J. Shroyer, E. Carlson, and D. Winter, whose expertise, comments, and support were invaluable.

Managing Editor: Michael Kane
Project Editor: Mali Apple
Production: The Cowans
Design: Detta Penna
Cover Art: Rachel Gage

ISBN 0-201-81525-7
Printed in the United States of America
   11 12 13 14   08 07 06 05

Dale Seymour Publications

Pearson Learning Group

1-800-321-3106
www.pearsonlearning.com

Contents

Introduction

These experiments are designed to increase students' understanding of some of the basic functions of advanced algebra and precalculus. Using either graphing calculators or computers, students become familiar with the behavior and basic properties of power, exponential, logarithmic, and rational functions.

Students gather data from a physical experiment. They plot the data, decide on a possible underlying function, and then experiment with parameters to determine whether the function can be made to approximate their data. Even if the data is "noisy" (inaccurate), students will be able to work with it, analyze it, and obtain a basic algebraic curve. Because the emphasis is on the underlying mathematics, it is unimportant whether a student's data leads to the "wrong" curve. Depending on the data, several functions may provide a reasonable fit.

In the optional extensions, students algebraically transform the problem to one of finding the line of best fit. The linearization step provides meaningful practice in manipulation.

Conducting the Experiments

The experiments can be used at two different levels. At the first level, they provide experience with the behavior of common nonlinear functions, among the basic functions of algebra: power functions, exponential and logarithmic functions, and rational functions. At the second level, students use algebraic manipulation to reduce the problem to a simpler case: fitting a straight line to the data. The manipulations chosen rely on a sense of the underlying function represented by the data. At both levels, students make use of technology. Points are plotted on a computer or calculator screen, and possible functions are graphed and adjusted. At the second level, a statistical tool is used to find the line of linear regression.

In each experiment, students select a "basic graph" from a menu of function graphs (from the Find the Equation worksheet). They then experiment, using technology rather than a ruler, to find the "best" such graph to represent their data. They assign and adjust values to two parameters. At the second level, students "linearize" their data. They then find the line of best fit for their linearized data, work the problem backwards, and compare it to the result obtained in the first level.

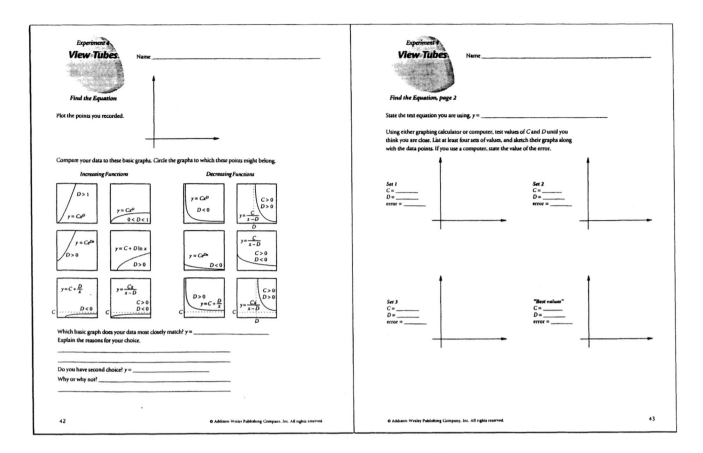

Plotting Points and Selecting the Basic Graph

For these experiments, it is preferable to plot points on a screen. It may be necessary to adjust the scale of the graph before deciding which basic function to try. Students look at the plotted points and ask themselves:

- Is it increasing or decreasing? Rapidly or slowly?
- Does it level off? Is there a horizontal asymptote?
- Is there a vertical asymptote?
- Does it cross the x-axis or the y-axis? Where?

Students then select a basic function to try. They are asked to indicate why they chose that function.

Experimenting to Find the Best Function

All the functions depend on the parameters C and D. Students must be able to graph the function on the same screen as the plotted points. This is most easily done using a computer program, but any graphing calculator can be programmed to allow for this. Sample computer and calculator programs are given in the appendices.

The Error

Most computer programs will include some measure of the mean square error, which allows students to judge whether changes in the trial values of C and D

are an improvement or not. Unfortunately, including computation of the error as part of a calculator program is not always feasible or possible.

Linearizing the Problem

Complete details are given below under the heading Experimental Analysis and Linearization Example. A word of warning, however! *Carry as many significant digits as possible. Rounding in the linearized problem can lead to large errors in the original problem.*

Experimental Analysis

Experimental analysis means students are to plot the data points and to select a possible basic function with which to work. Using a graphing calculator or computer, students adjust the constants C and D until the curve matches the data as closely as possible.

Experimental Analysis with a Graphing Calculator

The general procedure is as follows:

1. Enter a program that asks for values of C and D and then graphs the function.
2. Plot the data using the statistical features of a calculator.
3. Decide on a test function that depends on the parameters C and D.
4. Modify the program to graph the function.
5. When the screen becomes cluttered, clear it, re-plot the data, and continue.

Specific instructions for the most popular graphing calculators follows.

The Texas Instruments TI-81 Graphing Calculator

To plot the data points, deactivate all functions y. Set the range according to the data.

After the points have been plotted, you will want to graph the test function. Suppose you want to try the function $y = Cx^D$. Following is a program, called CURVE, to graph this function. Enter it as Program 1.

```
Prgm1:CURVE
Lbl 0
Disp "C"
Input C
Disp "D"
Input D
DrawF Cx^D        (replace with any basic equation)
Pause
Goto 0
```

Select **STAT**

Select **DATA ClrStat**

Select **STAT DATA**

Enter the data points, then press (**QUIT**)

Select **STAT DRAW Scatter**

Execute program CURVE

When the screen becomes cluttered:

Press (**ON**) (**CLEAR**) until the program stops

Select **DRAW ClrDraw**

Select **STAT DRAW Scatter** to replot the data

Execute program CURVE again

The Casio fx-7700G Graphing Calculator

Before entering the data, enter the appropriate version of the program below (adjust it according to the desired test function) in the COMP and WRT modes. (Pressing (**MODE**) (**÷**) puts you in the COMP [computer] mode, and pressing (**MODE**) (**2**) puts you in the WRT [writing] mode.) For example, suppose you want to try the function $y = Cx^D$. The program would be:

Lbl 0

"C": ?->C

"D": ?->D

Graph y = CX^D   *(replace with any basic equation)*

Goto 0

Press (**MODE**) (**1**) (**AC**) (to return to the run mode)

The program is easily modified after the data have been entered. To plot the data points, first set the appropriate range for your data, then enter

(**MODE**) (**÷**)

(**MODE**) (**SHIFT**) (**1**)

(**MODE**) (**SHIFT**) (**3**)

(**F2**) (**F3**) (**F1**)

Enter the first data point; for example, (3, 4): (**3**) (**SHIFT**) (**→**) (**4**) (**F1**)

You will see the plotted point. Continue entering points. When all the data have been entered, press (**G ←→ T**) (which toggles between the graphics screen and the text screen) to see the text screen. Then press (**MODE**) (**÷**) and run the program (press (**SHIFT**) (**RANGE**) (**F3**) the program number (**EXE**)) to graph the test function.

The Casio fx-7000G and fx-7700G Graphing Calculators

Select the linear regression mode. The data points cannot be stored; if the screen becomes cluttered, you will have to re-enter them.

The Casio fx-6300G Graphic Scientific Calculator

Before entering the data, enter the appropriate version of the program below (adjust it according to the desired test function) in the COMP and WRT modes. (Pressing (MODE) (+) puts you in the COMP [computer] mode, and pressing (MODE) (2) puts you in the WRT [writing] mode.) For example, suppose you want to try the function $y = Cx^D$. The program would be:

Lbl 0:"C": ?->C:"D": ?->D:Graph y = CX^D ▲ Goto 0 ▲

Press (MODE)(1)(AC) (to return to the run mode)

To plot the data points, first set the appropriate range for your data, then enter

(SHIFT) (MODE) (÷)

(SHIFT) (SCL) (EXE)

Enter the first data point; for example, (3, 4): (3) (SHIFT) (→) (4) (F1)

You will see the plotted point. Continue entering points. When all the data have been entered, press (MODE)(1) and run the program.

Experimental Analysis with a Computer

Although analysis can be done with a graphing calculator, most students will prefer to use a computer, both for ease of adjusting the parameters and the possibility of making hard copy. Most statistical computer programs will print out an error term. The most useful error indication is the *mean square error, E,* where:

$$E^2 = \sum (y_k - f(x_k))^2,$$

y_k is the y-coordinate of the kth data point, and $f(x_k)$ is the value of the approximating function. This is actually the sum of the squares of the errors; it should be as small as possible.

PC or Compatible Computers

The PC program Twiddle is excellent for investigating the fitting of curves to data. To request a copy, send a blank disk and three dollars to cover expenses to Mathematical Software, Department of Mathematics, University of Arizona, Tucson, Arizona 85721.

Apple II and Macintosh Computers

A listing of a BASIC program for each machine will be found in the appendices. Many other public domain and proprietary programs for statistical analysis are available.

Linearization Techniques

Given an equation in the variables x and y, find an equivalent linear equation in the variables T and W.

The equations of each of the basic graphs can be transformed to a linear equation. The new equation is the *transformed equation*. The variables T and W are the *transformed variables*. The linear regression function on a calculator can be used to find the best line through the transformed data points.

Basic Equation	Linearization Steps	Algebraic Explanation
$y = Cx^D$	$y = Cx^D$	
	$\ln y = \ln (Cx^D)$	take the natural log of both sides
	$\ln y = \ln C + \ln x^D$	simplify
	$\ln y = \ln C + D\ln x$	
		substitute $T = \ln x$ and $W = \ln y$,
	$W = A + BT$	resulting in a linear function
$y = Ce^{Dx}$	$y = Ce^{Dx}$	
	$\ln y = \ln (Ce^{Dx})$	take the natural log of both sides
	$\ln y = \ln C + Dx$	simplify
		substitute $T = x$ and $W = \ln y$,
	$W = A + BT$	resulting in a linear function
$y = C + D\ln x$	$y = C + D\ln x$	
		substitute $T = \ln x$ and $W = y$,
	$W = A + BT$	resulting in a linear function
$y = C + \dfrac{D}{x}$	$y = C + \dfrac{D}{x}$	
		substitute $T = \dfrac{1}{x}$ and $W = y$,
	$W = A + BT$	resulting in a linear function
$y = \dfrac{C}{x - D}$	$y = \dfrac{C}{x - D}$	
	$(x - D)y = C$	cross multiply
	$Dy = xy - C$	simplify
	$y = \dfrac{xy}{D} - \dfrac{C}{D}$	isolate y
		substitute $T = xy$ and $W = y$,
	$W = A + BT$	resulting in a linear function
$y = \dfrac{Cx}{x - D}$	$y = \dfrac{Cx}{x - D}$	
	$xy - Dy = Cx$	cross multiply
	$xy = Dy + Cx$	simplify
	$y = D\left(\dfrac{y}{x}\right) + C$	isolate y
		substitute $T = \dfrac{y}{x}$ and $W = y$,
	$W = A + BT$	resulting in a linear function

Your students may think of other ways to obtain a linear equation.

Many calculators will automatically find the best values of C and D for some of the basic functions, namely $y = Cx^D$, $y = C + D \ln x$, and $y = Ce^{Dx}$ (described as power, logarithmic, and exponential regressions). Not all data falls into one of these categories. However, a student who only uses the built-in functions and regards the best r value is likely to miss both. Students must understand how to apply the linearization techniques to all equations with two parameters.

The Line of Best Fit (Linear Regression)

Every set of data has a *line of best fit*. The word *best* means "better than any other"; it does not mean the line is a good fit. This line, also called the *line of linear regression*, usually goes between the data points rather than through them. The equation of this line is always given as:

$$y = A + Bx,$$

where A is the y-intercept and B is the slope. This labeling is used by all calculators with a statistical mode. **Important:** Students will need frequent reminding that the calculator finds $y = A + Bx$, not $y = Ax + B$.

What Makes $y = A + Bx$ the "Best" Line?

Suppose the experiment has resulted in five data points to be plotted, (x_1, y_1), (x_2, y_2), (x_3, y_3), (x_4, y_4), and (x_5, y_5), and that a line $y = C + Dx$ will be used to predict other values of y.

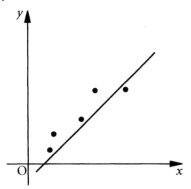

If the line is used instead of the points, $(C + Dx_1 - y_1)$ is the error in the prediction at the first point. The square of this error is $(C + Dx_1 - y_1)^2$. The sum of all the squares of the errors, called E^2, is given by the equation

$$E^2 = (C + Dx_1 - y_1)^2 + (C + Dx_2 - y_2)^2 + \ldots + (C + Dx_5 - y_5)^2.$$

The values of A and B produced by the calculator are the values of C and D that minimize E^2. Remember, the best line may not be very good.

What Is r, the Correlation Coefficient?

Study these graphs. Each graphs shows five data points, the line of best fit, and the value of the correlation coefficient *r*.

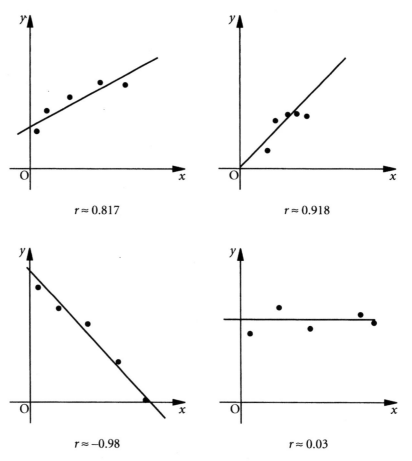

Acceptable Values for r

If $|r| > 0.98$, then the data probably represents a linear function. If $|r| < 0.98$, then either the function was not linear, or the data is "noisy"; that is, it contains errors.

Remember, the purpose of the experiments is to investigate nonlinear functions. Students must be consistent and take reasonable care when collecting data. If this has been done, students should decide on a basic function and proceed. If the *r* value is not acceptable, student should try a different basic curve, or redo the experiment.

How to Find A, B, and r on a Graphing Calculator

The Texas Instruments TI-81 Graphing Calculator

Select **STAT DATA ClrStat**

Select **STAT DATA Edit**

Enter the data points, then press (**2nd**)(**QUIT**)

For linear regression, select **STAT LinReg**

The regression coefficients *A*, *B*, and *r* will be displayed.

The Casio fx-6300G *and* fx-7000G *Graphing Calculators*

Press (MODE) (÷)

Press (SHIFT) (AC) [the (SCL) key] (EXE)

Enter the data. To enter the point (3, 4), press (3) (SHIFT) (( ) [the (,) key]
(4) (DT) [the $\boxed{x\sqrt{}}$ key]

To see the values of A, B, and r, press (SHIFT) (7) (EXE) (SHIFT) (8) (EXE) (SHIFT)
(9) (EXE)

The Casio fx-7700G *Graphing Calculator*

Press (MODE) (÷)

Press (MODE) (4)

Press (MODE) (SHIFT) (1) (MODE) (SHIFT) (4)

Press (F2) (F3) (F1)

Enter the data. To enter the point (3, 4), press (8) (SHIFT) (( ) [the (,) key]
(4) (F1)

When all the data are entered, press (F6)

To see the values of A, B, and r, press (F1) (EXE) (F2) (EXE) (F3) (EXE)

Experimental Analysis and Linearization Example

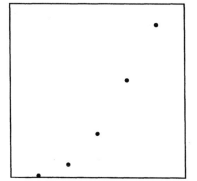

After leading the class through the following example, we suggest you discuss
and demonstrate how to transform some of the other equations. For example,
the following points arose from one of the experiments:

x	y
1	12
2	49
3	153
4	335
5	524

A look at the basic graphs suggests $y = Cx^D$.

Experimental Analysis

The following trials attempt to fit $y = Cx^D$ to the sample points.

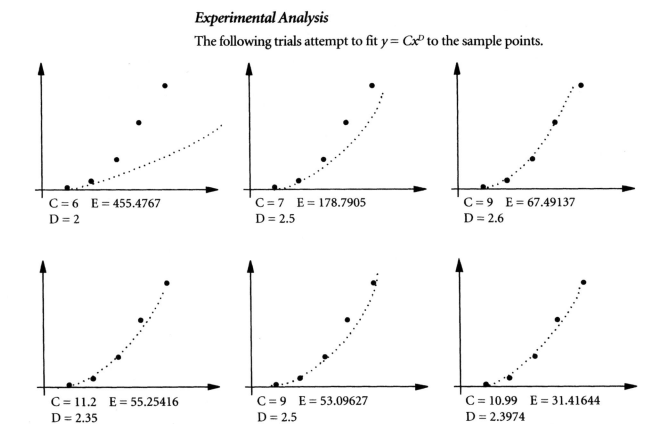

C = 6   E = 455.4767
D = 2

C = 7   E = 178.7905
D = 2.5

C = 9   E = 67.49137
D = 2.6

C = 11.2   E = 55.25416
D = 2.35

C = 9   E = 53.09627
D = 2.5

C = 10.99   E = 31.41644
D = 2.3974

Linearization Analysis

Taking the logarithms of both sides of $y = Cx^D$ gives:

$$\ln y = \ln C + D \ln x.$$

Let $T = \ln x$ and $W = \ln y$. The equation becomes:

$$W = (\ln C) + DT,$$

a linear equation. If we plot the points:

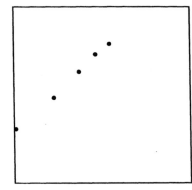

$T = \ln x$	$W = \ln y$
0	2.48
0.69	3.89
1.10	5.03
1.39	5.81
1.61	6.26

they appear collinear. Using a calculator and linear regression, the best linear approximation is

$$W = 2.3973 + 2.3974T,$$

with r ≈ 0.998.

Using algebra, we have:

$$\ln y = 2.3973 + 2.3974 \ln x$$

$$\ln y = \ln (e^{2.3973}) + \ln x^{2.3974}$$

$$\ln y = \ln 10.99 + \ln x^{2.3974}$$

10

$$y = 10.99\, x^{2.3974}$$

$$y \approx 11x^{2.4}$$

Using the original points and a power regression function, we obtain $A = 11.008$, $B = 2.3998$, and $r = 0.997$, which translates to $y = 11.008\, x^{2.3998}$.

The Experiments and the NCTM *Standards*

The Curriculum and Evaluation Standards for School Mathematics (copyright 1989 by the National Council of Teachers of Mathematics) provides a vision of mathematical literacy for a changing world and establishes guidelines to help revise mathematics curriculum to reflect the new definition of mathematical literacy. *Algebra Experiments* reflects the same spirit and addresses the same aims by implementing the goals espoused in the *Standards:* life-long learning, mathematically literate workers, and opportunity for all students. Learning to value mathematics, becoming confident mathematical problem-solvers, and learning to communicate and reason mathematically are outcomes that can be attained through the use of the experiments.

Changes in instructional practices in grades 9 to 12 mathematics required increased attention to the following goals:

- The active involvement of students in constructing and applying mathematical ideas
- Problem solving as a means as well as a goal of instruction
- The use of a variety of instructional formats (small groups, individual explorations, peer instruction, whole-class discussions, project work)
- The use of calculators and computers for tools for learning and doing mathematics
- Student communication of mathematical ideas orally and in writing

Changes in content in 9 to 12 mathematics require increased attention to the following topics:

- The use of real-world problems to motivate and apply algebraic theory.
- Integration of function concepts across topics at all grade levels
- The connections among a problem situation, its model as a function in symbolic form, and the graph of that function
- Function equations expressed in standardized form as checks on the reasonableness of graphs produced by graphing utilities
- Functions that are constructed as models of real-world problems
- Statistics

Cooperative Learning and Individual Accountability

Algebra Experiments combines the benefits of cooperative learning and individual responsibility. There is student interaction, discussion, informal peer instruction, shared ownership of a completed project, along with satisfaction and pride in a personally completed project and the individual's responsibility for understanding the mathematics and being able to complete the work involved.

Cooperative Learning

Most of the experiments require two people to successfully accomplish the collection of data. Even in those experiments one student could complete alone, two investigators help ensure accurate (consistent) data. During data collection, students will frequently make and share mathematical observations, such as: "Every time we add three, it goes up five. It must be linear."

After collecting data together, partners have a real interest in making the calculator or computer find a function that fits their data. There will be discussions and informal peer teaching: "I think it's levelling off—marbles can't roll forever." "We shouldn't use that point—the car nearly bounced." "Don't change C and D at the same time. We can't tell which is making the difference."

The Extension questions also produce discussions and conjectures to be challenged and defended: "I don't think that will happen. Let's try it and see." "Maybe it increases because . . ."

Individual Accountability

Each student should have a complete record of each experiment. Each should be able to describe in writing and with diagrams the exact nature of the experiment. Each should carry out the steps involved in finding a function and in answering the questions. It is important that each student write up the experiment completely, for purposes of evaluation, for possible inclusion in a portfolio, and as a record of accomplishment.

Evaluation and Assessment

The methods of evaluating student performance on an experiment and of assessing student understanding and learning will depend on how the experiment is used. An experiment might be used for enrichment, as an introduction, for motivation, for reinforcement, or as an adjunct to a mathematics course. An experiment may be marked as an assignment and returned, or it may be placed in a student's portfolio. Some suggestions for evaluation and examples of assessment questions follow.

Evaluating an Experiment: Marking Suggestions

Each experiment is divided naturally into three parts: Collect the Data (write-up and data collection), Find the Equation (determining a function representing the data), and Interpret the Data (responding to follow-up questions). How each part is counted depends on how the experiment is used. One method of evaluation, of course, is to give each part equal weight. Here are suggestions for awarding points to each part:

Collect the Data. If one assumes that the write-up, data collection, and graphing of the data are worth 10 points, the following is a possible allocation of credit:

- 2 points for a written description of the experiment
- 2 points for a diagram that shows the independent and dependent variables
- 2 points for identification of the variables and units
- 1 point for relevant identification and measurements of equipment
- 3 points for collecting and graphing the data

This distribution emphasizes the importance of understanding the experiment and being able to communicate, both visually and verbally, with others. The diagram may be minimal, but must clearly indicate the precise nature of the independent and dependent variables as well as the physical set-up. As long as the data produces points of a linear nature (which will be obvious from the graph), the precision of the data should not be unduly scrutinized. However, students should question an inconsistent point and perhaps repeat the experiment for that value of the independent variable. The physical limitations of the experiment should be also be respected; some values of the independent variable are too small or too large for the data to be meaningful.

Find the Equation. Give points for the reasons for the choice of basic graph, persistence in varying parameters, success in finding a "good" function if the problem is linearized, and for algebraic substitution and subsequent manipulation.

Interpret the Data. Give points for the substitution (manipulation-type) questions. Give points for inferential questions, such as, "How would the graph change if . . . ?" or "How would the equation change if . . . ?" The responses to these semi-open questions can indicate the depth of understanding of the concepts and relationships.

Assessment Alternatives and the Experiments

The present thinking about mathematics assessment is that both the product and the process should be assessed. The experimental write-up can be regarded as the product; the demonstrated ability to respond to the open-ended extension questions is also a product. The process includes the cooperative work in obtaining the data, as well as the behavior exhibited by students as they reason and respond to the questions.

Assessing the Product: Inclusion in a Mathematics Portfolio

The experiments are well suited for inclusion in a portfolio. Completion of the Collect the Data worksheet shows that the student understands the experiment and can communicate that understanding both in writing and with a diagram. Completion of the Find the Equation worksheet demonstrates possession of basic algebra skills and knowledge of their appropriate use. Performance on the Interpret the Data worksheet provides evidence of thinking beyond mechanical manipulations.

You might ask students who have done several experiments to select one or two write-ups for inclusion in their portfolios. They should write an introduction to the selected experiment(s) to explain their choice, the mathematics they used, and the significance of the constants C and D.

You should write a summary page, indicating for each part whether the student's work is exemplary, satisfactory, or inadequate. Include comments on the student's level of understanding.

Assessing the Process: Observation

As students work on the experiments, you can observe each group and answer these questions:

> Are both partners focused on the experiment?
>
> Is each student concentrating?
>
> Are they working independently, as well as together?
>
> What is the nature of their conversations; are they exchanging facts and discussing results?
>
> What level of understanding do the discussions show?
>
> Are they observing patterns?

Note any interesting or informative student comments.

Assessing the Process: Questions

Four basic types of questions can be used to test students' understanding. One or two examples of each are given here. The Teaching Notes pages for many of the experiments suggest others. Extension questions can also be easily adapted.

Experiment-based Questions

Chris did one of the experiments, but beyond writing x and y forgot to identify the variables. Here is the data he obtained.

Which basic function(s) would you try with this data? Why?

x	y
7	46
9	54
11	58
13	61
15	63
17	65
19	66
20	68

Graph-based Questions

1. Here is the graph of $y = C + \dfrac{D}{x}$, with $C > 0$ and $D > 0$. Sketch the graph of $y = C - \dfrac{D}{x}$ for the same values of C and D.

2. You did the Rising Damp experiment and obtained this graph.

You did it again with more absorbant blotting paper. Sketch what the graph for the new data might look like.

If your original curve was $y = 2x^{0.3}$, which would you expect to change more, the 2 or the 0.3? Why?

Equation-based Questions

1. For the Flat, Black, and Circular experiment, Si found $y = 2.6x^{1.9}$ while Sam found $y = 5.3x^{2.1}$. Can you tell who used miniature marshmallows and who used marbles?

2. Janelle put a pan of cold water on the stove and turned on the burner. Before the water boiled, she had collected her data. Using time as the independent variable, she measured the water temperature every 30 seconds. She found that $y = 60 \, e^{0.0038x}$. What was the temperature of the water when she started? How long did it take the water to reach 120° F?

Open-ended (What-if) Questions

Aretha did an experiment and is sure that her graph belongs to the family $y = \sqrt{(Cx^3 + D)}$. Suppose she's guessed correctly. What change of variables to W and T would result in a making W a linear function of T?

Teaching the Experiments

All experiments follow the same format.

Introduction—Whole Class

- Present the experiment: "If I change the diameter of the marble, will it roll further?"

- Demonstrate the procedure: "I'll drop it from a height of 90 cm and measure." "It bounced 51 cm."

- Identify the independent and dependent variables: "What do I decide? What's the independent variable?" "What result do I measure? What's the dependent variable?"

- Identify the experimental constants: "Will all the balls bounce the same? The results depend on which ball. We need to note that."

Write-up and Drawing—Individual Work

Before beginning the experiment, each student should complete the drawing and description of the experiment on the Collect the Data worksheet. The drawing should clearly show x and y, the independent and dependent variables. Where appropriate, equipment should be identified and measured.

Data Collection—Students Work in Pairs

Have students form cooperative learning groups of two (or three) students. Have each student in the group record the data on his or her own Collect the Data worksheet.

A test value of the independent variable should be chosen and the experiment tried. Once consistent results have been obtained, data collection should begin in earnest.

In some experiments, it is wise to repeat the measurements two or three times and take the "best" value. In some cases, this will be the middle value of y. In others, it will the maximum value or the average value.

All data should be entered into the Data Collection boxes. When the collection is completed, the points to be plotted should be transferred to the Points to Be Graphed chart on the right side of the page.

Sample student work. Please note: student answers will vary.

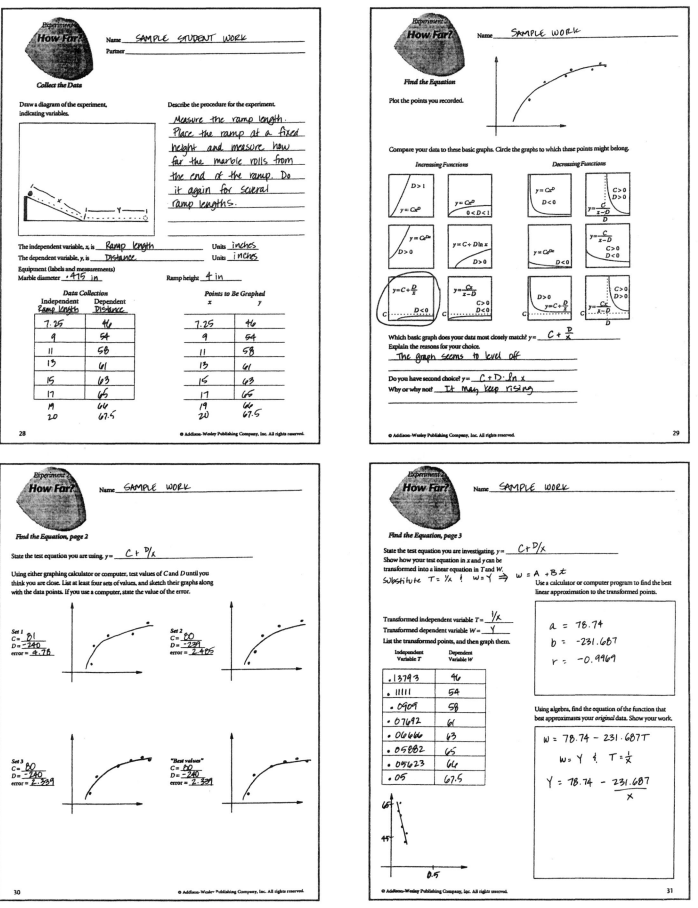

Experiment 2
How Far?   Name SAMPLE STUDENT WORK

Partner _____

Collect the Data

Draw a diagram of the experiment, indicating variables.

Describe the procedure for the experiment.

Measure the ramp length. Place the ramp at a fixed height and measure how far the marble rolls from the end of the ramp. Do it again for several ramp lengths.

The independent variable, x, is _Ramp length_   Units _inches_
The dependent variable, y, is _Distance_   Units _inches_

Equipment (labels and measurements)
Marble diameter _.415 in_   Ramp height _4 in_

Data Collection		Points to Be Graphed	
Independent Ramp length	Dependent Distance	x	y
7.25	46	7.25	46
9	54	9	54
11	58	11	58
13	61	13	61
15	63	15	63
17	65	17	65
19	66	19	66
20	67.5	20	67.5

28   © Addison-Wesley Publishing Company, Inc. All rights reserved.

Experiment 2
How Far?   Name SAMPLE WORK

Find the Equation

Plot the points you recorded.

Compare your data to these basic graphs. Circle the graphs to which these points might belong.

Increasing Functions   *Decreasing Functions*

Which basic graph does your data most closely match? $y = C + \dfrac{D}{x}$
Explain the reasons for your choice.

The graph seems to level off

Do you have second choice? $y = C + D \cdot \ln x$
Why or why not? It may keep rising

© Addison-Wesley Publishing Company, Inc. All rights reserved.   29

Experiment 2
How Far?   Name SAMPLE WORK

Find the Equation, page 2

State the test equation you are using, $y = C + \dfrac{D}{x}$

Using either graphing calculator or computer, test values of C and D until you think you are close. List at least four sets of values, and sketch their graphs along with the data points. If you use a computer, state the value of the error.

Set 1
C = 81
D = -240
error = 4.78

Set 2
C = 80
D = -239
error = 2.406

Set 3
C = 80
D = -240
error = 2.339

"Best values"
C = 80
D = -240
error = 2.339

30   © Addison-Wesley Publishing Company, Inc. All rights reserved.

Experiment 2
How Far?   Name SAMPLE WORK

Find the Equation, page 3

State the test equation you are investigating. $y = C + \dfrac{D}{x}$
Show how your test equation in x and y can be transformed into a linear equation in T and W.

Substitute $T = \dfrac{1}{x}$ & $w = Y \Rightarrow w = A + Bt$

Use a calculator or computer program to find the best linear approximation to the transformed points.

Transformed independent variable $T = \dfrac{1}{x}$
Transformed dependent variable $W = Y$

List the transformed points, and then graph them.

Independent Variable T	Dependent Variable W
.13793	46
.11111	54
.0909	58
.07692	61
.06666	63
.05882	65
.05623	66
.05	67.5

$a = 78.74$
$b = -231.687$
$r = -0.9969$

Using algebra, find the equation of the function that best approximates your *original* data. Show your work.

$w = 78.74 - 231.687T$

$w = Y$ & $T = \dfrac{1}{x}$

$Y = 78.74 - \dfrac{231.687}{x}$

© Addison-Wesley Publishing Company, Inc. All rights reserved.   31

Name SAMPLE WORK

Interpret the Data

Copy your final equation here. $y =$ $C + D/x$

Answer the following questions. Show your work.

1. Rewrite the equation to express distance as a function of ramp length.

   distance $= 78.74 - 231.687/$ ramp length

2. If you were to use the entire length traveled, including the ramp, what effect do you think it would have on your curve?

   The curve would be steeper at the beginning

   Predict how C and D would change. C will increase D will decrease

3. Solve your equation for x.
   $$Y = C + D/x$$
   $$xY = xC + D$$
   $$xY - xC = D$$
   $$x(Y-C) = D$$

   $x = \frac{D}{Y-C}$

4. Write the new equation, expressing the ramp length (the independent variable) as a function of distance traveled (the dependent variable).

   Ramp length $= \frac{-231.687}{distance - 78.74}$ or $\frac{231.687}{78.74 - distance}$

5. Use a graphing calculator to graph both functions on the same set of axes. Copy the graphs below. Indicate the x and y ranges.

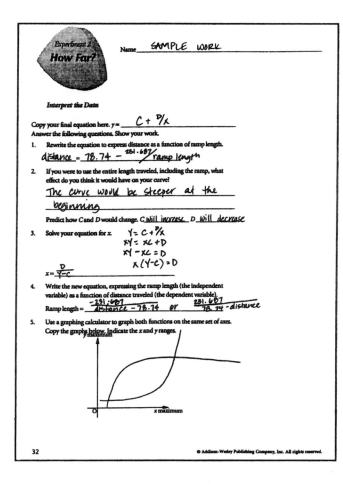

y maximum

x maximum

List of the Experiments

	Experiment	Most Common Function	Independent Variable	Dependent Variable
1	Bouncing Ball	$y = C + D \ln x$	drop height	time until third bounce
2	How Far?	rational	ramp length	roll distance
3	Rolling Stone	$y = Cx^D$	marble diameter	roll distance
4	View Tubes	rational	distance to object	viewable vertical distance
5	Focusing	rational	distance from lens to object	distance from lens to image
6	Swingtime	$y = Cx^D$	pendulum length	pendulum period
7	Flat, Black, and Circular	$y = Cx^D$	lid diameter	number of objects needed to encircle rim
8	Filling Funnels	$y = Cx^D$	fill depth	number of objects needed to fill funnel
9	Cool Down	$y = Ce^{Dx}$	time	temperature
10	The Rising Damp	$y = Ce^{Dx}$	mark number	time
11	Mirror, Mirror on the Floor	rational	mark height	distance from mirror
12	Musical Glasses	$y = Ce^{Dx}$	note number	water level
13	Balancing Act	rational	distance from pivot hole	number of weights
14	Falling Marbles	$y = Cx^D$	ramp height	roll distance

Experiment 1

Bouncing Ball

Teaching Notes

In this experiment, the elapsed time until the third bounce of a ball is a function of the drop height of the ball. The drop height of the ball is the *independent variable,* and the elapsed time from the drop of the ball until the third bounce is the *dependent variable.* This experiment doesn't need much introduction, but measuring the timing of the bounce takes practice.

Equipment

assorted balls (golf balls, table tennis balls, super balls), 1 per group

> *Number the balls. Choose balls that make noise as they bounce, and use at least two of each type.*

yardsticks, 1 per group

stopwatches, or watches that display seconds, 1 per group

graph paper, 1 sheet per student

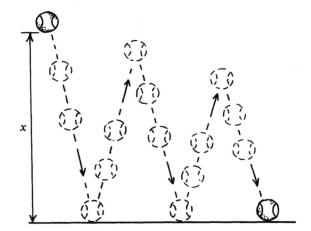

Procedure

Students work in groups of two. The first student drops a ball from a given height (the drop height); the second student measures elapsed time from the drop of the ball to the third bounce.

Have partners practice taking measurements until they are getting consistent results. If, when their points are plotted, the group finds that one point is badly out of line, they should redo the experiment from the same drop height.

Organizing and Analyzing Class Results

Write the students' equations on the board. Ask students to identify the equation that represents golf balls, table tennis balls, etc.

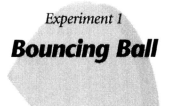

Experiment 1

Bouncing Ball

Teaching Notes, page 2

Extension

This experiment can be done as class demonstration with the help of 3 or 4 students. Have each student plot the data and sketch the curve.

To investigate exponential decay, drop a ball from the ceiling or other high place and measure the height of the bounce. Then, drop the ball from the height it attained on the previous bounce. In this situation, the *independent variable* is the bounce number, and the *dependent variable* is the height reached after bounce *x*.

Say: Chris and Tyler used the same ball as in their Bouncing Ball linear experiment. Their equation was $y = 0.59x + 2$. Does that agree with our new data? Is each bounce about 0.59 times as high as the previous bounce? If that's true, after a bounce the height should be about $(0.59)^2$ times the original height.

Plot the data and the graph of $y = (0.59)^x$ on a graphing calculator to see whether $y = (0.59)^x$ is a good approximation to the data (it will be close). Have students use the equation from their original experiment to plot height as a function of bounce number for their ball.

Experiment 1

Bouncing Ball

Name _____

Partner _____

Collect the Data

Draw a diagram of the experiment, indicating variables.

Describe the procedure for the experiment.

_____

_____

_____

_____

_____

_____

_____

_____

_____

_____

The independent variable, x, is _____ Units _____

The dependent variable, y, is _____ Units _____

Equipment (labels and measurements)

Type of ball _____ Number _____

Data Collection			Points to Be Graphed	
Independent	Dependent		x	y
_____	_____			

Experiment 1

Bouncing Ball

Name_____

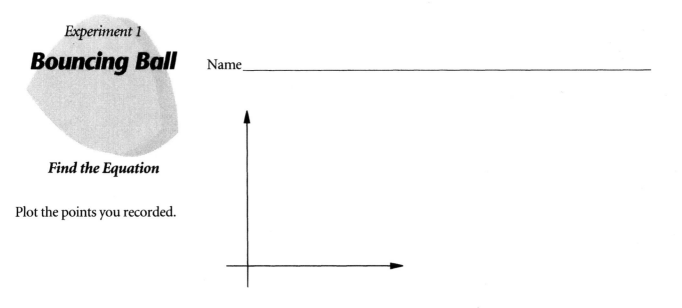

Find the Equation

Plot the points you recorded.

Compare your data to these basic graphs. Circle the graphs to which these points might belong.

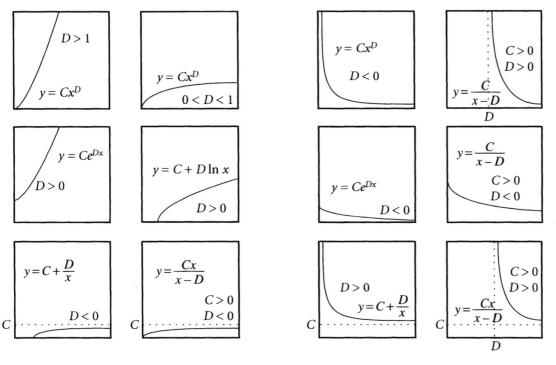

Which basic graph does your data most closely match? $y =$ _____
Explain the reasons for your choice.

_____

_____

Do you have second choice? $y =$ _____

Why or why not? _____

_____

Bouncing Ball

Name _____

Find the Equation, page 2

State the test equation you are using. $y = $ _____

Using either graphing calculator or computer, test values of C and D until you think you are close. List at least four sets of values, and sketch their graphs along with the data points. If you use a computer, state the value of the error.

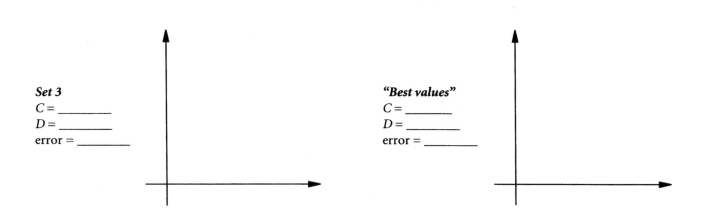

Set 1
$C = $ _____
$D = $ _____
error = _____

Set 2
$C = $ _____
$D = $ _____
error = _____

Set 3
$C = $ _____
$D = $ _____
error = _____

"Best values"
$C = $ _____
$D = $ _____
error = _____

Experiment 1

Bouncing Ball

Name _____

Find the Equation, page 3

State the test equation you are investigating. $y =$ _____

Show how your test equation in x and y can be transformed into a linear equation in T and W.

Use a calculator or computer program to find the best linear approximation to the transformed points.

Transformed independent variable $T =$ _____

Transformed dependent variable $W =$ _____

List the transformed points, and then graph them.

Independent Variable T	Dependent Variable W

Using algebra, find the equation of the function that best approximates your *original* data. Show your work.

Experiment 1

Bouncing Ball

Name _____

Interpret the Data

Copy your final equation here. $y =$ _____

Answer the following questions. Show your work.

1. Rewrite the equation to express the elapsed time as a function of the drop height.

    _____ = _____

2. If you were to use a basketball, what effect do you think it would have on your curve?

    _____

    _____

    Predict how C and D would change. C_____ D_____

3. Solve your equation for x.

    $x =$ _____

4. Write the new equation, expressing the drop height (the independent variable) as a function of elapsed time (the dependent variable).

    Drop height = _____

5. Use a graphing calculator to graph both functions on the same set of axes. Copy the graphs below. Indicate the x and y ranges.

26

In this experiment, the distance a marble rolls from the end of a ramp is a function of the length of the ramp. The ramp length is the *independent variable,* and the distance the marble rolls from the end of the ramp is the *dependent variable.* The height of the ramp is kept constant.

Equipment

marbles, 1 per group

ramps of varying lengths, 15"–22"

> *Number the ramps. If they are made from slats of wood, bevel the downhill end. Vinyl gutters are inexpensive; most building-supply stores will cut a 10-foot length for you.*

blocks, books, or another material to raise the ramps

yardsticks, 1 per group

graph paper, 1 sheet per student

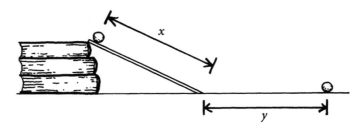

Procedure

The floor must be carpeted for the marble to stop rolling after a reasonable distance.

Students choose a fixed value for the height of the ramp. Then, for a given ramp length (the independent variable), they measure the distance the marble rolls from the end of the ramp (the dependent variable). The "best" measurement in this experiment will be the average or middle distance for the given ramp length.

Have students start the marble at the high end of the ramp. The ramp should be high enough to induce rolling, but not so steep as to cause the marble to bounce. A 15-inch ramp should not be raised more than 6 inches.

Experiment 2

How Far?

Name _____

Partner _____

Collect the Data

Draw a diagram of the experiment, indicating variables.

[ ]

Describe the procedure for the experiment.

_____

_____

_____

_____

_____

_____

_____

_____

_____

_____

_____

The independent variable, x, is _____   Units _____

The dependent variable, y, is _____   Units _____

Equipment (labels and measurements)

Marble diameter _____   Ramp height _____

| *Data Collection* | | *Points to Be Graphed* | |
Independent	Dependent	x	y

28

Experiment 2

How Far?

Name_____

Find the Equation

Plot the points you recorded.

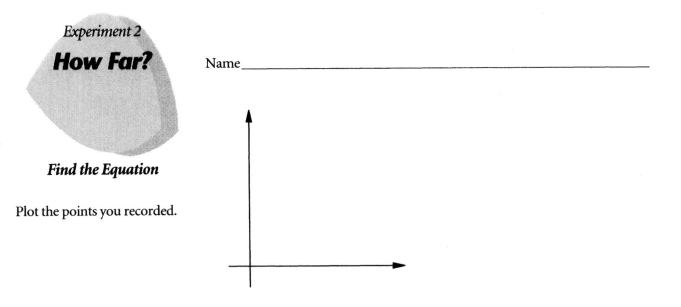

Compare your data to these basic graphs. Circle the graphs to which these points might belong.

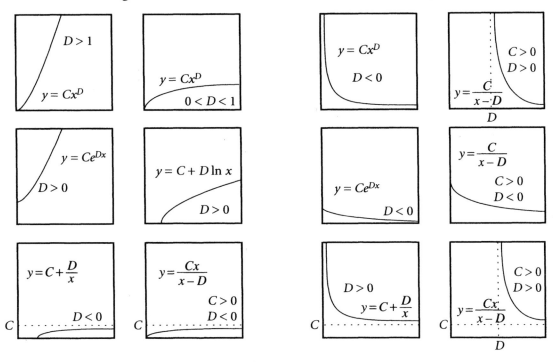

Increasing Functions *Decreasing Functions*

$D > 1$

$y = Cx^D$

$y = Cx^D$
$0 < D < 1$

$y = Ce^{Dx}$
$D > 0$

$y = C + D \ln x$
$D > 0$

$y = C + \dfrac{D}{x}$
$D < 0$

$y = \dfrac{Cx}{x - D}$
$C > 0$
$D < 0$

$y = Cx^D$
$D < 0$

$y = \dfrac{C}{x - D}$
$C > 0$
$D > 0$

$y = \dfrac{C}{x - D}$
$C > 0$
$D < 0$

$D > 0$
$y = C + \dfrac{D}{x}$

$y = \dfrac{Cx}{x - D}$
$C > 0$
$D > 0$

Which basic graph does your data most closely match? $y =$ _____
Explain the reasons for your choice.

_____

_____

Do you have second choice? $y =$ _____

Why or why not? _____

_____

Experiment 2

How Far?

Name _____

Find the Equation, page 2

State the test equation you are using. $y =$ _____

Using either graphing calculator or computer, test values of C and D until you think you are close. List at least four sets of values, and sketch their graphs along with the data points. If you use a computer, state the value of the error.

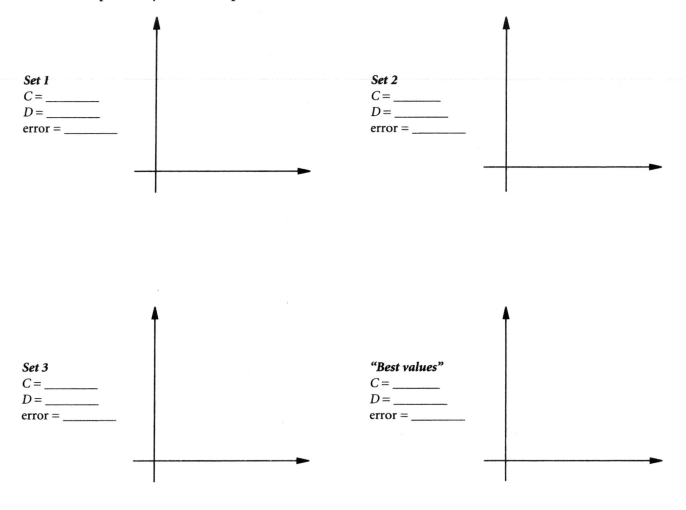

Set 1
$C =$ _____
$D =$ _____
error = _____

Set 2
$C =$ _____
$D =$ _____
error = _____

Set 3
$C =$ _____
$D =$ _____
error = _____

"Best values"
$C =$ _____
$D =$ _____
error = _____

Experiment 2

How Far?

Name _____

Find the Equation, page 3

State the test equation you are investigating. $y = $ _____

Show how your test equation in x and y can be transformed into a linear equation in T and W.

Use a calculator or computer program to find the best linear approximation to the transformed points.

Transformed independent variable $T = $ _____

Transformed dependent variable $W = $ _____

List the transformed points, and then graph them.

Independent Variable T	Dependent Variable W

Using algebra, find the equation of the function that best approximates your *original* data. Show your work.

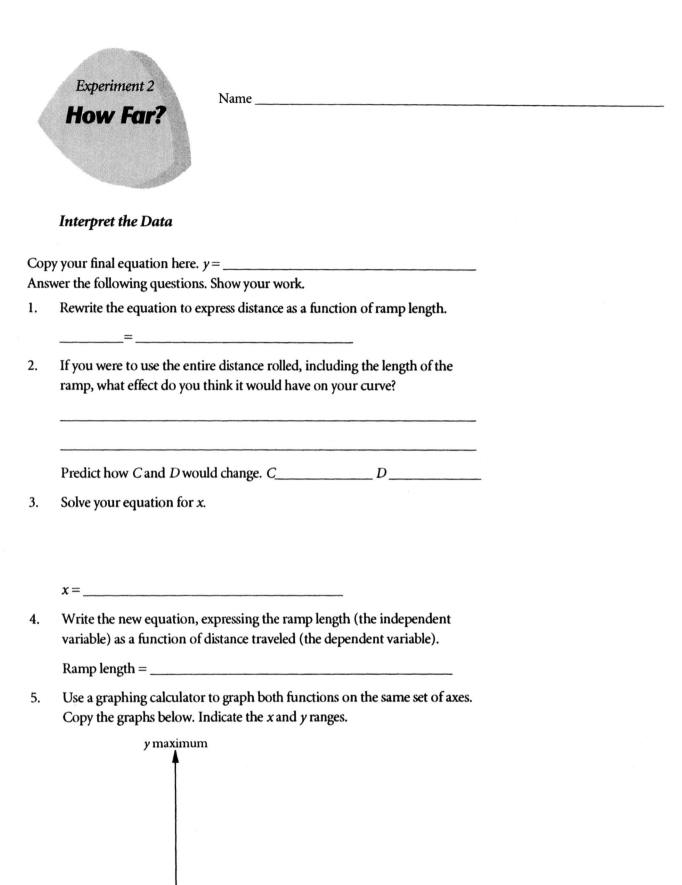

Experiment 2

How Far?

Name _____

Interpret the Data

Copy your final equation here. $y =$ _____

Answer the following questions. Show your work.

1. Rewrite the equation to express distance as a function of ramp length.

    _____ = _____

2. If you were to use the entire distance rolled, including the length of the ramp, what effect do you think it would have on your curve?

    _____

    _____

    Predict how C and D would change. C _____ D _____

3. Solve your equation for x.

    $x =$ _____

4. Write the new equation, expressing the ramp length (the independent variable) as a function of distance traveled (the dependent variable).

    Ramp length = _____

5. Use a graphing calculator to graph both functions on the same set of axes. Copy the graphs below. Indicate the x and y ranges.

    y maximum

    O   *x* maximum

In this experiment, the distance rolled from the end of a ramp is a function of the diameter of a marble. The diameter of the marble is the *independent variable,* and the distance the marble rolls from the end of the ramp is the *dependent variable.*

Equipment

marbles of differing diameters, 5 or 6 per group

carpeted floor

ramps of varying lengths, 15"–22"

Number the ramps. If they are made from slats of wood, bevel the downhill end. Vinyl gutters are inexpensive; most building-supply stores will cut up a 10-foot length for you.

yardsticks, 1 per group

calipers

blocks, books, or another material to raise the ramps

graph paper, 1 sheet per student

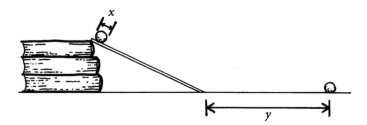

Procedure

The floor must be carpeted for the marble to stop rolling after a reasonable distance.

Students choose a fixed value for the height of the ramp and a fixed ramp length. They measure the diameter of the marble, the independent variable. It is easier to measure the diameter of the marble with calipers. Then students roll the marble down the ramp and measure the distance the marble rolls from the end of the ramp, the dependent variable. The "best" measurement in this experiment will be the average or middle distance for the given diameter.

Have students start the marble at the high end of the ramp. The ramp should be high enough to induce rolling, but not so steep as to cause the marble to bounce. A 15-inch ramp should not be raised more than 6 inches.

Experiment 3

Rolling Stone

Name _____

Partner _____

Collect the Data

Draw a diagram of the experiment, indicating variables.

Describe the procedure for the experiment.

_____

_____

_____

_____

_____

_____

_____

_____

_____

_____

_____

The independent variable, x, is _____   Units _____

The dependent variable, y, is _____   Units _____

Equipment (labels and measurements)

Ramp number _____   Ramp length _____

Data Collection

Independent	Dependent

Points to Be Graphed

x	y

Rolling Stone

Name_____

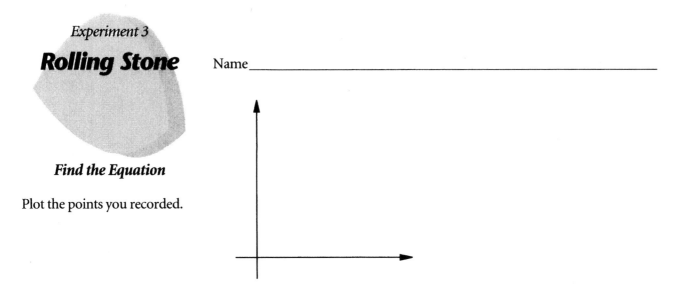

Find the Equation

Plot the points you recorded.

Compare your data to these basic graphs. Circle the graphs to which these points might belong.

Increasing Functions			*Decreasing Functions*	

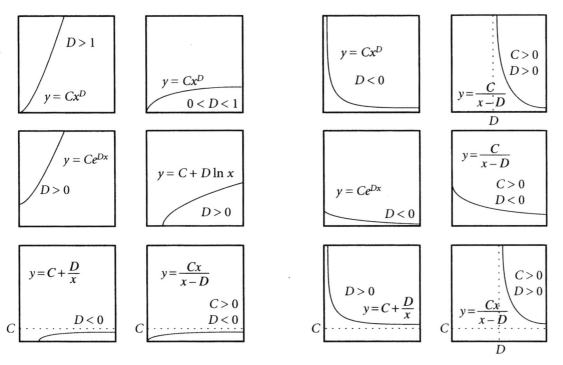

Which basic graph does your data most closely match? $y =$ _____
Explain the reasons for your choice.

_____

_____

Do you have second choice? $y =$ _____

Why or why not? _____

_____

Expriment 3

Rolling Stone

Name _____

Find the Equation, page 2

State the test equation you are using. $y =$ _____

Using either graphing calculator or computer, test values of C and D until you
think you are close. List at least four sets of values, and sketch their graphs along
with the data points. If you use a computer, state the value of the error.

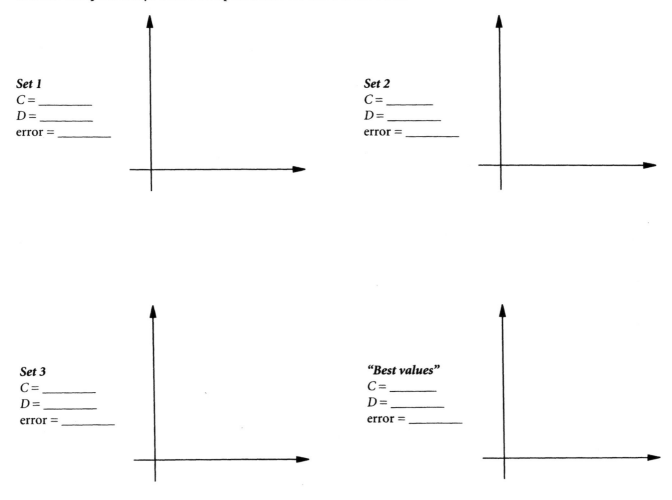

Set 1
$C =$ _____
$D =$ _____
error = _____

Set 2
$C =$ _____
$D =$ _____
error = _____

Set 3
$C =$ _____
$D =$ _____
error = _____

"Best values"
$C =$ _____
$D =$ _____
error = _____

Experiment 3

Rolling Stone

Name _____

Find the Equation, page 3

State the test equation you are investigating. $y =$ _____
Show how your test equation in x and y can be
transformed into a linear equation in T and W.

Use a calculator or computer program to find the best
linear approximation to the transformed points.

Transformed independent variable $T =$ _____

Transformed dependent variable $W =$ _____

List the transformed points, and then graph them.

Independent Variable T	Dependent Variable W

Using algebra, find the equation of the function that
best approximates your *original* data. Show your work.

Rolling Stone

Name _____

Interpret the Data

Copy your final equation here. $y =$ _____
Answer the following questions. Show your work.

1. Rewrite the equation to express the distance rolled as a function of the diameter of the marble.

    _____ = _____

2. If you were to use the entire distance rolled (including the length of the ramp), what effect do you think it would have on your curve?

    _____

    _____

    Predict how C and D would change. C_____ D_____

3. Solve your first equation for x.

    $x =$ _____

4. Write the new equation, expressing the marble diameter (the independent variable) as a function of distance rolled (the dependent variable).

    Marble diameter = _____

5. Use a graphing calculator to graph both functions on the same set of axes. Copy the graphs below. Indicate the x and y ranges.

View Tubes

Teaching Notes

In this experiment, the viewable vertical distance is a function of the length of a tube. The length of the tube is the *independent variable*, and the viewable vertical distance is the *dependent variable*. The diameter of the tube and the distance from the wall remain constant.

Equipment

assorted tubes, 1 per group

> *Number the tubes. Try to have several tubes of the same length but different diameters, and some of the same diameter but different lengths. Tubes from kitchen wraps (plastic, wax paper, foil) work well. Sawed-off lengths from the core of a carpet or fabric bolt add variety. Cut them long enough so that rolling the eye doesn't distort the measurements.*

yardsticks, 2 per group

graph paper, 1 sheet per student

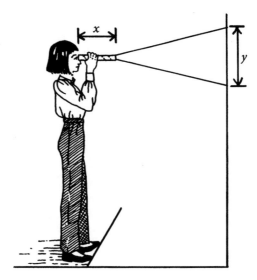

Procedure

Students work in groups of two or three. Students wearing glasses should not be paired with students who do not.

Both students should record their viewable vertical distance for each value of *x*. The measured values should be averaged for plotting. Reasonable accuracy is obtained if students measure the distance from the toes. In theory, the intercept is the diameter of the tube; rough measurement may lead to a straight line with a negative intercept.

Students will have to decide on units. (Measuring the distance from the wall in feet and the viewable vertical distance in inches works well. Some students, however, will not want to use different units for the two variables).

Have students obtain all their points first, then decide how to scale their graphs. If their *x*-measurements are 4, 5, 6, and 7 (that is, close together), they should allow 5 squares to a unit on the graph.

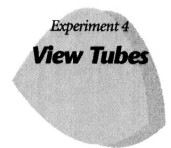
After recording the measurements of the length and diameter of their tubes, students stand a fixed distance from the wall and measure the viewable vertical distance.

All students should record their viewable vertical distance for each value of x and average the measured values for plotting.

Organizing and Analyzing Class Results

The viewable vertical distance is nonlinear function of the tube length. Your students may be interested in the geometric analysis.

l: distance from the end of tube to the wall

x: length of the tube

d: diameter of the tube

y: viewable vertical distance

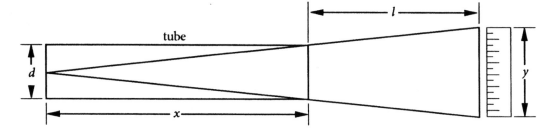

Using similar triangles with d and l held constant,

$$\frac{\frac{y}{2}}{l+x} = \frac{\frac{d}{2}}{x}$$

$$x\left(\frac{y}{2}\right) = \frac{d}{2}(l+x)$$

$$y = \frac{d(l+x)}{x}$$

or

$$y = \frac{dl}{x} + d$$

Experiment 4

View Tubes

Name _____

Partner _____

Collect the Data

Draw a diagram of the experiment, indicating variables.

Describe the procedure for the experiment.

The independent variable, x, is _____ Units _____

The dependent variable, y, is _____ Units _____

Equipment (labels and measurements)

Distance from wall _____   Fixed diameter of tubes _____

| *Data Collection* | | *Points to Be Graphed* | |
Independent	Dependent	x	y

View Tubes

Name_____

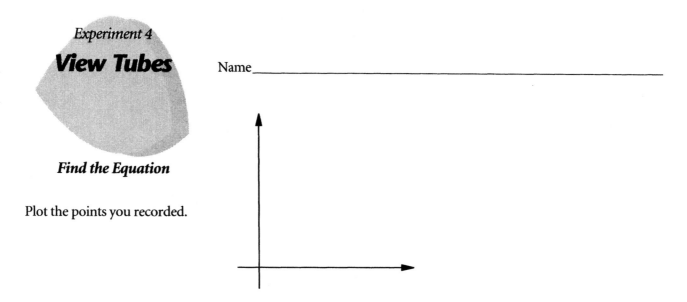

Find the Equation

Plot the points you recorded.

Compare your data to these basic graphs. Circle the graphs to which these points might belong.

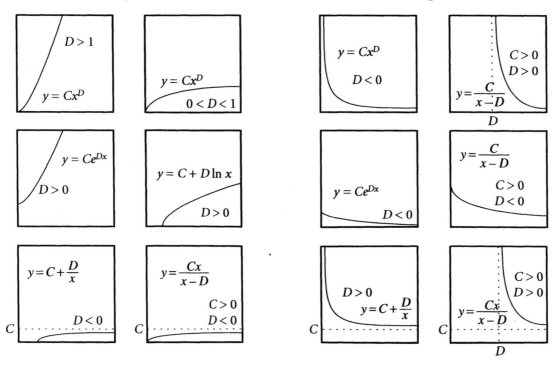

Increasing Functions

$D > 1$
$y = Cx^D$

$y = Cx^D$
$0 < D < 1$

$y = Ce^{Dx}$
$D > 0$

$y = C + D \ln x$
$D > 0$

$y = C + \dfrac{D}{x}$
C $D < 0$

$y = \dfrac{Cx}{x - D}$
$C > 0$
$D < 0$
C

Decreasing Functions

$y = Cx^D$
$D < 0$

$C > 0$
$D > 0$
$y = \dfrac{C}{x - D}$
D

$y = \dfrac{C}{x - D}$
$C > 0$
$D < 0$
$y = Ce^{Dx}$
$D < 0$

$D > 0$
$y = C + \dfrac{D}{x}$
C

$C > 0$
$D > 0$
$y = \dfrac{Cx}{x - D}$
C
D

Which basic graph does your data most closely match? $y =$ _____
Explain the reasons for your choice.

_____

_____

Do you have second choice? $y =$ _____

Why or why not? _____

_____

View Tubes

Name _____

Find the Equation, page 2

State the test equation you are using. $y =$ _____

Using either graphing calculator or computer, test values of C and D until you think you are close. List at least four sets of values, and sketch their graphs along with the data points. If you use a computer, state the value of the error.

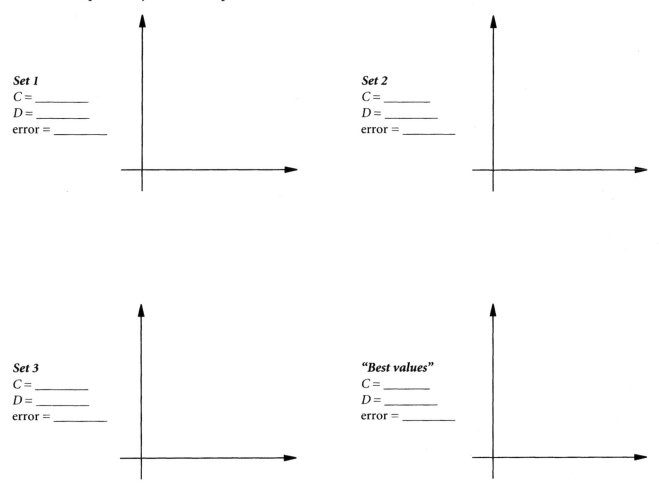

Set 1
$C =$ _____
$D =$ _____
error = _____

Set 2
$C =$ _____
$D =$ _____
error = _____

Set 3
$C =$ _____
$D =$ _____
error = _____

"Best values"
$C =$ _____
$D =$ _____
error = _____

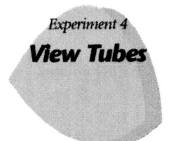

Experiment 4

View Tubes

Name _____

Find the Equation, page 3

State the test equation you are investigating. $y =$ _____

Show how your test equation in x and y can be transformed into a linear equation in T and W.

Use a calculator or computer program to find the best linear approximation to the transformed points.

Transformed independent variable $T =$ _____

Transformed dependent variable $W =$ _____

List the transformed points, and then graph them.

Independent Variable T	Dependent Variable W

Using algebra, find the equation of the function that best approximates your *original* data. Show your work.

Experiment 4

View Tubes

Name _____

Interpret the Data

Copy your final equation here. $y =$ _____

Answer the following questions. Show your work.

1. Rewrite the equation to express the viewable vertical distance as a function of the length of the tube.

    _____ = _____

2. If you stood further away from the wall, what effect do you think it would have on your curve?

    _____

    _____

    Predict how C and D would change. C _____ D _____

3. Solve your first equation for x.

    $x =$ _____

4. Write the new equation, expressing the length of the tube (the independent variable) as a function of viewable vertical distance (the dependent variable).

    Tube length = _____

5. Use a graphing calculator to graph both functions on the same set of axes. Copy the graphs below. Indicate the x and y ranges.

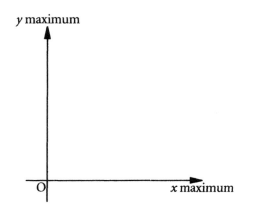

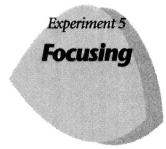

Experiment 5
Focusing

Teaching Notes

In this experiment, the distance between an image screen and a lens is a function of the distance between the lens and an object. The distance between the lens and the object is the *independent variable*, and the distance between the image screen and the lens for the object to be in focus is the *dependent variable*.

Equipment

yardsticks, 1 per group

magnifying glass or lens, 1 per group

flashlight, 1 per group

plain cardboard, 6" × 6", 2 pieces per group

graph paper, 1 sheet per student

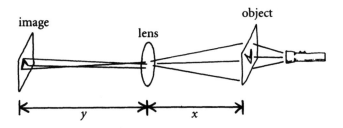

Procedure

Have students set up the following objects in a straight line, perpendicular to the surface, in a slightly darkened room: the image screen (cardboard), the magnifying glass, the object (cardboard with the V-shaped cutout), and the flashlight.

Beginning with around 12 inches for the distance between the lens and the object, have students shine the flashlight through the V-shaped cutout directed at the lens. (The minimum distance from the object to the lens will depend on the lens.) Then, have them move the image screen away from the lens until the is in sharp focus.

Because it is difficult to get the image focused precisely, have students do several trials and average their results.

Experiment 5

Focusing

Name _____

Partner _____

Collect the Data

Draw a diagram of the experiment, indicating variables.

Describe the procedure for the experiment.

_____

_____

_____

_____

_____

_____

_____

_____

_____

_____

The independent variable, x, is _____ Units _____

The dependent variable, y, is _____ Units _____

Equipment (labels and measurements)

Lens _____

Data Collection	
Independent	Dependent
_____	_____

Points to Be Graphed	
x	y

Experiment 5

Focusing

Name_____

Find the Equation

Plot the points you recorded.

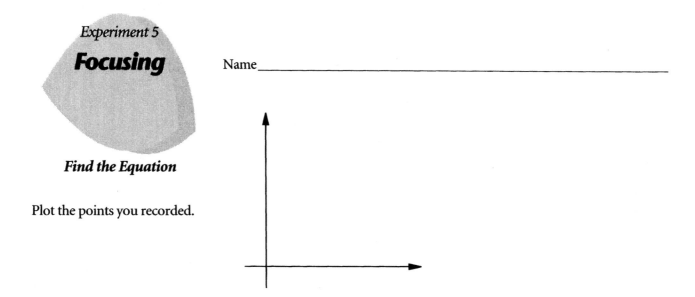

Compare your data to these basic graphs. Circle the graphs to which these points might belong.

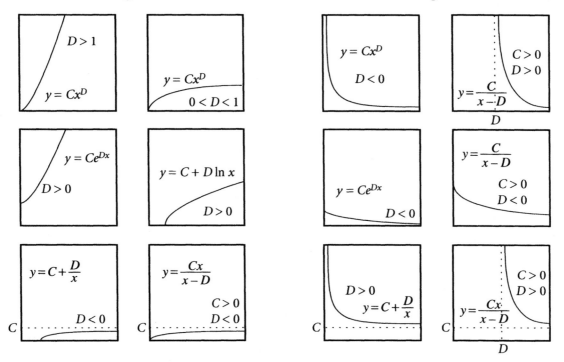

Increasing Functions   *Decreasing Functions*

$D > 1$

$y = Cx^D$

$y = Cx^D$
$0 < D < 1$

$y = Cx^D$
$D < 0$

$C > 0$
$D > 0$
$y = \dfrac{C}{x - D}$

D

$y = Ce^{Dx}$

$D > 0$

$y = C + D \ln x$

$D > 0$

$y = Ce^{Dx}$
$D < 0$

$y = \dfrac{C}{x - D}$
$C > 0$
$D < 0$

$y = C + \dfrac{D}{x}$

C
$D < 0$

$y = \dfrac{Cx}{x - D}$
$C > 0$
$D < 0$

C

$D > 0$
$y = C + \dfrac{D}{x}$

C

$C > 0$
$D > 0$
$y = \dfrac{Cx}{x - D}$

C

D

Which basic graph does your data most closely match? $y =$ _____
Explain the reasons for your choice.

_____

_____

Do you have second choice? $y =$ _____

Why or why not? _____

_____

Experiment 5

Focusing

Name _____

Find the Equation, page 2

State the test equation you are using. $y =$ _____

Using either graphing calculator or computer, test values of C and D until you think you are close. List at least four sets of values, and sketch their graphs along with the data points. If you use a computer, state the value of the error.

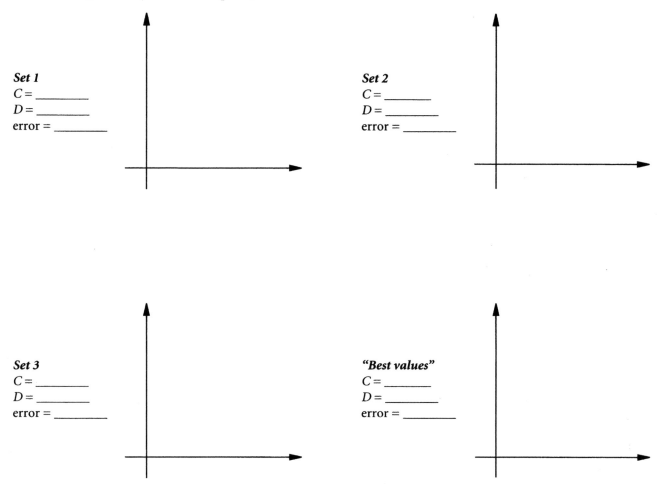

Set 1
$C =$ _____
$D =$ _____
error = _____

Set 2
$C =$ _____
$D =$ _____
error = _____

Set 3
$C =$ _____
$D =$ _____
error = _____

"Best values"
$C =$ _____
$D =$ _____
error = _____

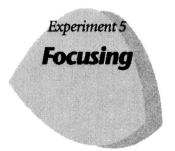

Experiment 5

Focusing

Name _____

Find the Equation, page 3

State the test equation you are investigating. $y =$ _____

Show how your test equation in x and y can be transformed into a linear equation in T and W.

Use a calculator or computer program to find the best linear approximation to the transformed points.

Transformed independent variable $T =$ _____

Transformed dependent variable $W =$ _____

List the transformed points, and then graph them.

Independent Variable T	Dependent Variable W

Using algebra, find the equation of the function that best approximates your *original* data. Show your work.

Experiment 5

Focusing

Name _____

Interpret the Data

Copy your final equation here. $y =$ _____

Answer the following questions. Show your work.

1. Rewrite the equation to express the distance of the lens to the screen as a
    function of the distance of the object to the lens.

    _____ = _____

2. Solve your first equation for x.

    $x =$ _____

3. Write the new equation, expressing the distance of the object to the lens
    (the independent variable) as a function of distance of the lens to the
    screen (the dependent variable).

    Object distance = _____

4. Use a graphing calculator to graph both functions on the same set of axes.
    Copy the graphs below. Indicate the x and y ranges.

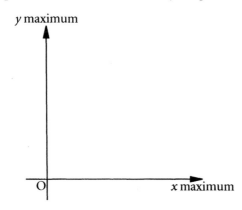

Experiment 6
Swingtime

Teaching Notes

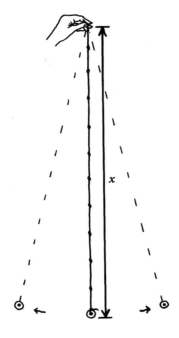

In this experiment, the time needed for one period is a function of the length of a pendulum. The length of the pendulum (the distance from the knot to the far end of the weight) is the *independent variable*, and the period of the pendulum (the time for one complete swing back and forth) is the *dependent variable*.

Equipment

lengths of string, about 4'

weights (heavy nuts or washers)

stopwatches, or watches that display seconds, 1 per group

yardsticks, 1 per group

graph paper, 1 sheet per student

Procedure

Students make pendulums by tying a weight to one end of the string. They tie six to eight knots in the string, one at least 36 inches from the weight and one about 8 inches from the weight.

To measure the period, students hold the string by one of the knots, extend the string, raise the weight, and release it. They measure the time necessary for 10 complete swings back and forth, and they divide by 10 to obtain the period. Students should time the swings carefully. Most values of the dependent variable will be between 1.1 and 1.8 seconds.

Students will find $y = Cx^D$.

Experiment 6

Swingtime

Name _____

Partner _____

Collect the Data

Draw a diagram of the experiment, indicating variables.

Describe the procedure for the experiment.

_____

_____

_____

_____

_____

_____

_____

_____

_____

_____

The independent variable, x, is _____ Units _____

The dependent variable, y, is _____ Units _____

<table>
<tr><th colspan="2">Data Collection</th><th colspan="2">Points to Be Graphed</th></tr>
<tr><th>Independent</th><th>Dependent</th><th>x</th><th>y</th></tr>
<tr><td></td><td></td><td></td><td></td></tr>
<tr><td></td><td></td><td></td><td></td></tr>
<tr><td></td><td></td><td></td><td></td></tr>
<tr><td></td><td></td><td></td><td></td></tr>
<tr><td></td><td></td><td></td><td></td></tr>
<tr><td></td><td></td><td></td><td></td></tr>
</table>

Experiment 6
Swingtime

Name_____

Find the Equation

Plot the points you recorded.

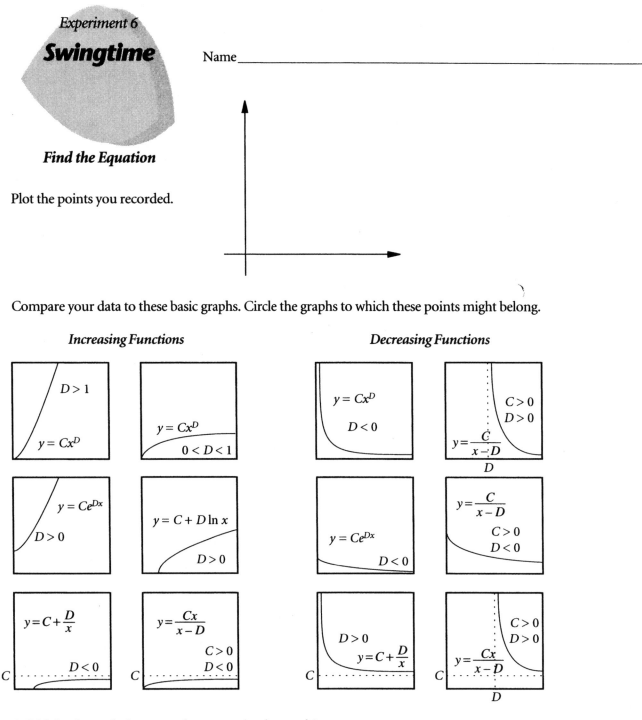

Compare your data to these basic graphs. Circle the graphs to which these points might belong.

Increasing Functions

$D > 1$

$y = Cx^D$

$y = Cx^D$

$0 < D < 1$

$y = Ce^{Dx}$

$D > 0$

$y = C + D \ln x$

$D > 0$

$y = C + \dfrac{D}{x}$

$D < 0$

C

$y = \dfrac{Cx}{x - D}$

$C > 0$

$D < 0$

C

Decreasing Functions

$y = Cx^D$

$D < 0$

$y = \dfrac{C}{x - D}$

$C > 0$

$D > 0$

D

$y = \dfrac{C}{x - D}$

$C > 0$

$D < 0$

$y = Ce^{Dx}$

$D < 0$

$D > 0$

$y = C + \dfrac{D}{x}$

C

$y = \dfrac{Cx}{x - D}$

$C > 0$

$D > 0$

C

D

Which basic graph does your data most closely match? $y =$ _____
Explain the reasons for your choice.

_____

_____

Do you have second choice? $y =$ _____

Why or why not? _____

_____

54

Experiment 6

Swingtime

Name _____

Find the Equation, page 2

State the test equation you are using. $y =$ _____

Using either graphing calculator or computer, test values of C and D until you think you are close. List at least four sets of values, and sketch their graphs along with the data points. If you use a computer, state the value of the error.

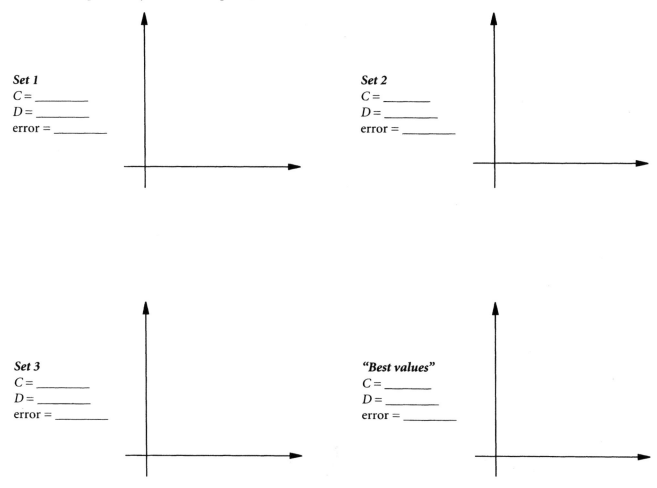

Set 1
$C =$ _____
$D =$ _____
error = _____

Set 2
$C =$ _____
$D =$ _____
error = _____

Set 3
$C =$ _____
$D =$ _____
error = _____

"Best values"
$C =$ _____
$D =$ _____
error = _____

Experiment 6

Swingtime

Name _____

Find the Equation, page 3

State the test equation you are investigating. $y =$ _____

Show how your test equation in x and y can be transformed into a linear equation in T and W.

Use a calculator or computer program to find the best linear approximation to the transformed points.

Transformed independent variable $T =$ _____

Transformed dependent variable $W =$ _____

List the transformed points, and then graph them.

Independent Variable T	Dependent Variable W

Using algebra, find the equation of the function that best approximates your *original* data. Show your work.

Experiment 6

Swingtime

Name _____

Interpret the Data

Copy your final equation here. $y =$ _____

Answer the following questions. Show your work.

1. Rewrite the equation to express the period of the pendulum as a function of the length of the pendulum.

    _____ = _____

2. What effect do you think a heavier weight would have on your curve?

    _____

    Predict how C and D would change. C _____ D _____

3. Solve your first equation for x.

    $x =$ _____

4. Write the new equation, expressing the length of the pendulum (the independent variable) as a function of period (the dependent variable).

    Pendulum length = _____

5. Use a graphing calculator to graph both functions on the same set of axes. Copy the graphs below. Indicate the x and y ranges.

Experiment 7

Flat, Black, and Circular

Teaching Notes

In this experiment, the number of pieces of cereal is a function of the diameter of a lid. The diameter of the lid is the *independent variable,* and the number of cereal pieces needed to fill the lid is the *dependent variable.*

Equipment

large supply of spherical-shaped breakfast cereal, dried chickpeas, or other dried beans

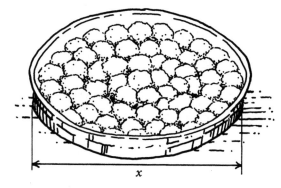

Pieces need not be identical. More than one type is desirable for the extension questions; label each type by letter.

assorted jar lids, 5 or 6 different sizes per group

Lids should be flat, 3" to 8" (7 cm to 20 cm) in diameter, and have a rim to contain the pieces; try lids for take-out drinks and plastic lids from coffee and peanut cans.

rulers, 1 per group

cups or containers to hold each group's supply of cereal or beans

graph paper, 1 sheet per student

Procedure

Have groups select a circle (lid) and measure its diameter by placing a ruler across the circle. Diameter may be measured in inches or centimeters, but a group should use the same units throughout the experiment.

Next, have them fill the lid with pieces of cereal (or beans), counting as they go.

Organizing and Analyzing Class Results

Students will have found their graph is of the form $y = Cx^D$.

Sort students' graphs according to whether the diameter was measured in centimeters or in inches. List the equations for each of the groups.

Ask questions such as the following: Can you tell from the equations alone what objects were used? If you were to use much larger circles and golf balls instead of cereal or beans, how might the equation change? What if you were to use unpopped popcorn? (From looking at the equations, students should deduce that C will change, but D will remain approximately 2.)

Now, sort the graphs according to the type of cereal or bean used. For those who used chickpeas, for example, *ask:* Who used centimeters? Who used inches? How can you tell from the graphs? From the equations?

Ask: If we were to use a much larger circle, like a drum from the school band, and measure the diameter in feet, how would the graph change? How would that affect the equation?

58

Experiment 7

Flat, Black, and Circular

Name _____

Partner _____

Collect the Data

Draw a diagram of the experiment, indicating variables.

Describe the procedure for the experiment.

The independent variable, x, is _____ Units _____

The dependent variable, y, is _____ Units _____

Equipment (labels and measurements)

Type of cereal _____

Lid numbers _____

Data Collection		Points to Be Graphed	
Independent	Dependent	x	y

Experiment 7
Flat, Black, and Circular

Name_____

Find the Equation

Plot the points you recorded.

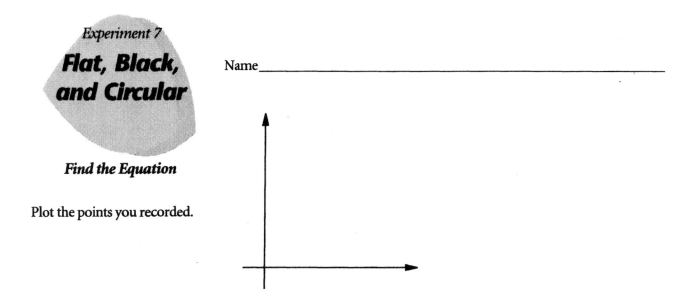

Compare your data to these basic graphs. Circle the graphs to which these points might belong.

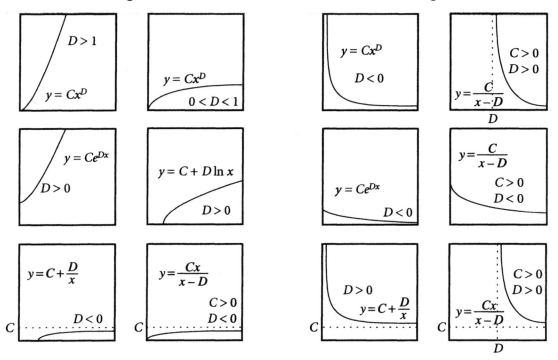

Increasing Functions   *Decreasing Functions*

$D > 1$
$y = Cx^D$

$y = Cx^D$
$0 < D < 1$

$y = Cx^D$
$D < 0$

$C > 0$
$D > 0$
$y = \dfrac{C}{x - D}$
D

$y = Ce^{Dx}$
$D > 0$

$y = C + D \ln x$
$D > 0$

$y = Ce^{Dx}$
$D < 0$

$y = \dfrac{C}{x - D}$
$C > 0$
$D < 0$

$y = C + \dfrac{D}{x}$
$D < 0$
C

$y = \dfrac{Cx}{x - D}$
$C > 0$
$D < 0$
C

$D > 0$
$y = C + \dfrac{D}{x}$
C

$C > 0$
$D > 0$
$y = \dfrac{Cx}{x - D}$
C
D

Which basic graph does your data most closely match? $y = $ _____
Explain the reasons for your choice.

_____

_____

Do you have second choice? $y = $ _____

Why or why not? _____

_____

Flat, Black, and Circular

Name _____

Find the Equation, page 2

State the test equation you are using. $y =$ _____

Using either graphing calculator or computer, test values of C and D until you think you are close. List at least four sets of values, and sketch their graphs along with the data points. If you use a computer, state the value of the error.

Set 1
$C =$ _____
$D =$ _____
error = _____

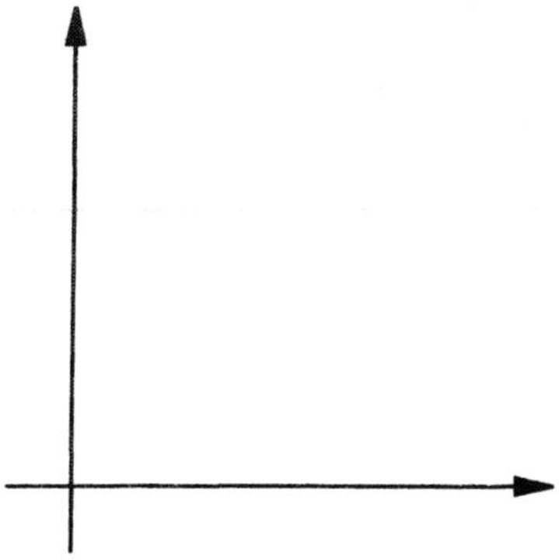

Set 2
$C =$ _____
$D =$ _____
error = _____

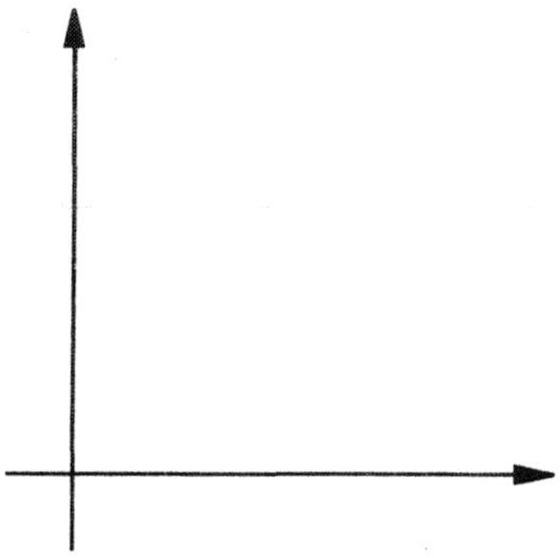

Set 3
$C =$ _____
$D =$ _____
error = _____

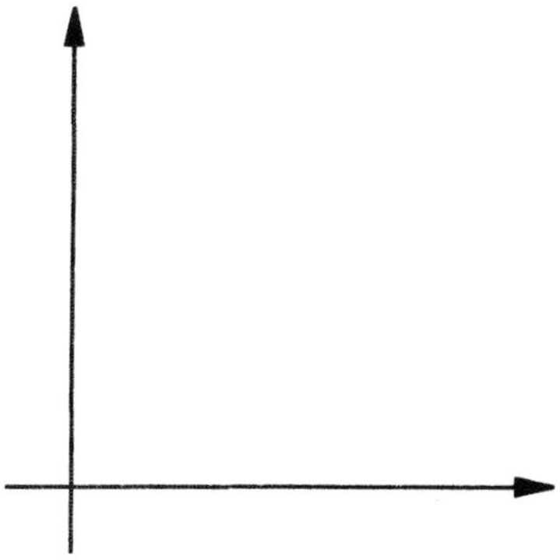

"Best values"
$C =$ _____
$D =$ _____
error = _____

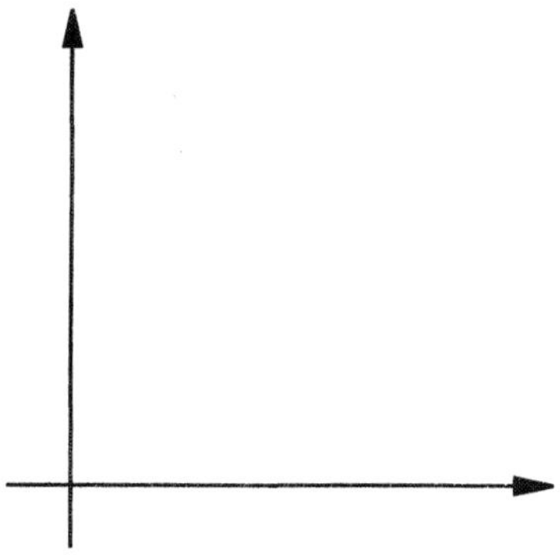

Experiment 7

Flat, Black, and Circular

Find the Equation, page 3

Name _____

State the test equation you are investigating. $y =$ _____
Show how your test equation in x and y can be
transformed into a linear equation in T and W.

Use a calculator or computer program to find the best
linear approximation to the transformed points.

Transformed independent variable $T =$ _____

Transformed dependent variable $W =$ _____

List the transformed points, and then graph them.

Independent Variable T	Dependent Variable W

Using algebra, find the equation of the function that
best approximates your *original* data. Show your work.

62

Experiment 7

Flat, Black, and Circular

Name _____

Interpret the Data

Copy your final equation here. $y =$ _____

Answer the following questions. Show your work.

1.   Rewrite the equation to express the number of cereal pieces as a function of the diameter of the lid.

   _____ = _____

2.   If you were to use unpopped popcorn kernels, what effect do you think it would have on your curve?

   _____

   _____

   Predict how C and D would change. C _____ D _____

3.   Solve your first equation for x.

   $x =$ _____

4.   Write the new equation, expressing the diameter (the independent variable) as a function of the number of cereal pieces (the dependent variable).

   Diameter = _____

5.   Use a graphing calculator to graph both functions on the same set of axes. Copy the graphs below. Indicate the x and y ranges.

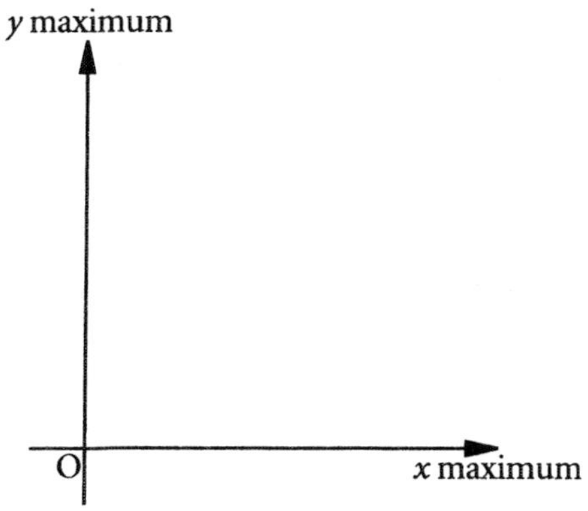

In this experiment, the number of objects needed to fill a funnel to a given depth is a function of the depth. The depth of the filling is the *independent variable*, and the number of objects needed to fill the funnel to a given depth is the *dependent variable*.

Equipment

funnels, 1 per group

> *The top part of 2-liter soda bottle works well; cut the bottle at the height where it starts to curve and invert. The cap makes a good base.*

large supply of large-size breakfast cereal, large dried beans, or other objects

> *Goldfish-shaped crackers work surprisingly well; to reach a depth of 4 inches requires approximately 200 goldfish-shaped crackers.*

overhead projector pen and ruler, 1 per group

graph paper, 1 sheet per student

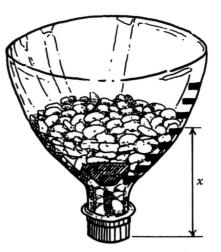

Procedure

Have students begin by marking depth levels, using the overhead projector pen and ruler, on the side the funnel. Starting at a depth of 1.5 inches, they should mark every half inch up to 5 inches.

Add enough filler to reach the first mark. The funnel should be shaken down frequently to level the contents. Emphasize that the independent variable is the vertical height (the depth), not the slant height.

Students will probably find their graph is of the form $y = Cx^D$.

For a true, right circular cone with constant radius/height ratio of $\frac{r}{h} = k$, volume is given by the equation:

$$V = \tfrac{1}{3}\pi r^2 h = \tfrac{1}{3}\pi k^2 h^2 h = (\text{constant}) \times h^3.$$

The result for the soda-bottle funnels will be nearly a cubic function.

Filling Funnels

Name _____

Partner _____

Collect the Data

Draw a diagram of the experiment, indicating variables.

Describe the procedure for the experiment.

_____

_____

_____

_____

_____

_____

_____

_____

_____

_____

_____

The independent variable, x, is _____   Units _____

The dependent variable, y, is _____   Units _____

Equipment (labels and measurements)

Type of cereal _____   Funnel _____

Data Collection	
Independent	Dependent

Points to Be Graphed	
x	y

Experiment 8
Filling Funnels

Name_____

Find the Equation

Plot the points you recorded.

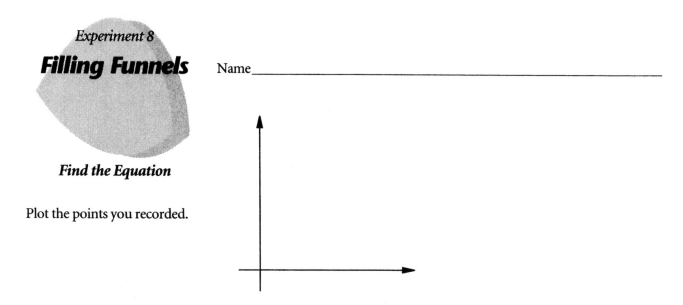

Compare your data to these basic graphs. Circle the graphs to which these points might belong.

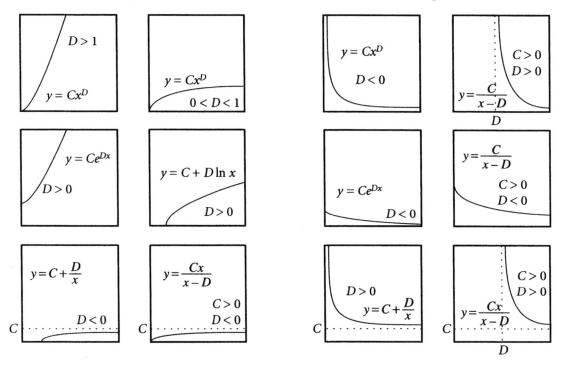

Which basic graph does your data most closely match? $y =$ _____
Explain the reasons for your choice.

_____

_____

Do you have second choice? $y =$ _____

Why or why not? _____

_____

66

Experiment 8

Filling Funnels

Name _____

Find the Equation, page 2

State the test equation you are using. $y =$ _____

Using either graphing calculator or computer, test values of C and D until you
think you are close. List at least four sets of values, and sketch their graphs along
with the data points. If you use a computer, state the value of the error.

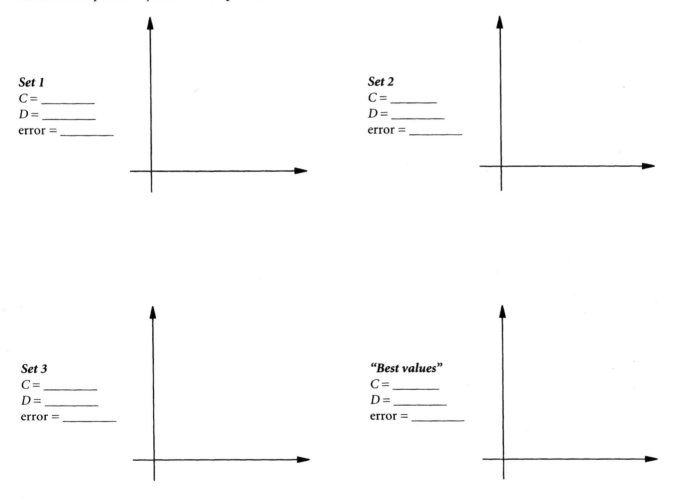

Set 1
$C =$ _____
$D =$ _____
error = _____

Set 2
$C =$ _____
$D =$ _____
error = _____

Set 3
$C =$ _____
$D =$ _____
error = _____

"Best values"
$C =$ _____
$D =$ _____
error = _____

Experiment 8
Filling Funnels

Name _____

Find the Equation, page 3

State the test equation you are investigating. $y =$ _____
Show how your test equation in x and y can be
transformed into a linear equation in T and W.

Use a calculator or computer program to find the best
linear approximation to the transformed points.

Transformed independent variable $T =$ _____

Transformed dependent variable $W =$ _____

List the transformed points, and then graph them.

Independent Variable T	Dependent Variable W

Using algebra, find the equation of the function that
best approximates your *original* data. Show your work.

68

Filling Funnels

Name _____

Interpret the Data

Copy your final equation here. $y = $ _____
Answer the following questions. Show your work.

1. Rewrite the equation to express the number of cereal pieces as a function
    of the depth of the filling.

    _____ = _____

2. If you were to use unpopped popcorn kernels, what effect do you think it
    would have on your curve?

    _____

    _____

    Predict how C and D would change. C_____ D _____

3. Solve your first equation for x.

    $x = $ _____

4. Write the new equation, expressing the depth of the filling (the indepen-
    dent variable) as a function of the number of cereal pieces (the dependent
    variable).

    Filling depth = _____

5. Use a graphing calculator to graph both functions on the same set of axes.
    Copy the graphs below. Indicate the x and y ranges.

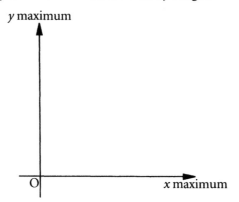

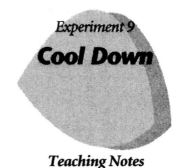

Experiment 9

Cool Down

Teaching Notes

In this experiment, the temperature of the water is a function of the elapsed time. The elapsed time is the *independent variable*, and the temperature of the water is the *dependent variable*.

Equipment

warm or hot water (not boiling)

thermometers, 1 per group

> *May be of the indoor or outdoor type, but should be vertical as they will be used to stir the water. Thermometers with degrees in Celsius are preferable.*

standard-size ice cubes

containers, 1 per group

> *Pint-size containers work well.*

stopwatches, or watches that display seconds, 1 per group

graph paper, 1 sheet per student

Procedure

In each group, one student handles the thermometer and calls the time, and the other student records the data. Have students pour about 1 cup of warm water into the container and add 3 or 4 ice cubes, starting the stopwatch at the same time. The thermometer handler should stir the mixture constantly. At various intervals, students should note the time and the temperature.

Although the independent variable is time, students will find it convenient to note the time at which the temperature reaches certain degree marks. The thermometer handler should call out each decrease of 5° C. The time recorder notes the total elapsed time.

Extension

Students will find their graphs are of the form $y = Ce^{Dx}$, where $D < 0$. Cool Down is an example of exponential decay.

Experiment 9

Cool Down

Name _____

Partner _____

Collect the Data

Draw a diagram of the experiment, indicating variables.

Describe the procedure for the experiment.

The independent variable, x, is _____ Units _____

The dependent variable, y, is _____ Units _____

Equipment (labels and measurements)

Data Collection			
Independent	Dependent
_____	_____

Points to Be Graphed	
x	y

Experiment 9

Cool Down

Name_____

Find the Equation

Plot the points you recorded.

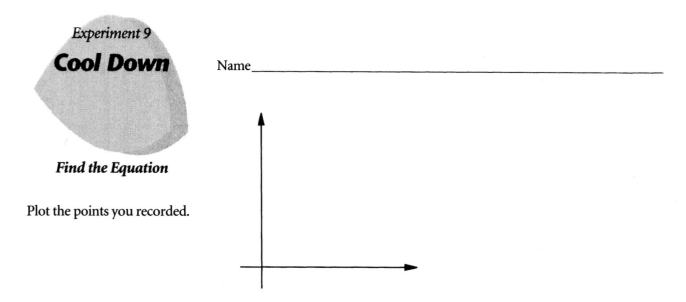

Compare your data to these basic graphs. Circle the graphs to which these points might belong.

| *Increasing Functions* | *Decreasing Functions* |

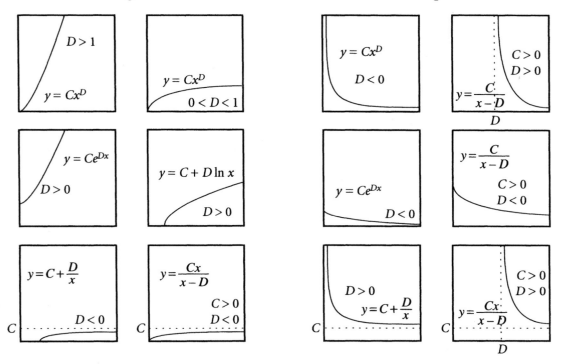

Which basic graph does your data most closely match? $y =$ _____
Explain the reasons for your choice.

_____

_____

Do you have second choice? $y =$ _____

Why or why not? _____

_____

72

Experiment 9

Cool Down

Name _____

Find the Equation, page 2

State the test equation you are using. $y =$ _____

Using either graphing calculator or computer, test values of C and D until you think you are close. List at least four sets of values, and sketch their graphs along with the data points. If you use a computer, state the value of the error.

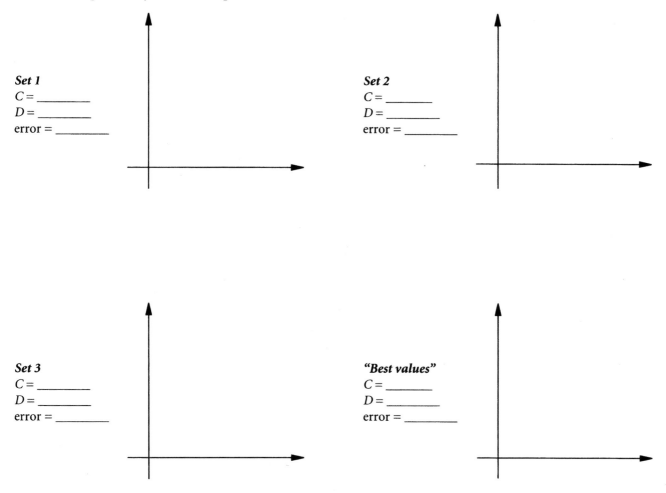

Set 1
$C =$ _____
$D =$ _____
error = _____

Set 2
$C =$ _____
$D =$ _____
error = _____

Set 3
$C =$ _____
$D =$ _____
error = _____

"Best values"
$C =$ _____
$D =$ _____
error = _____

Experiment 9
Cool Down

Name _____

Find the Equation, page 3

State the test equation you are investigating. $y =$ _____
Show how your test equation in x and y can be
transformed into a linear equation in T and W.

Use a calculator or computer program to find the best
linear approximation to the transformed points.

Transformed independent variable $T =$ _____

Transformed dependent variable $W =$ _____

List the transformed points, and then graph them.

Independent Variable T	Dependent Variable W

Using algebra, find the equation of the function that
best approximates your *original* data. Show your work.

74

Experiment 9
Cool Down

Name _____

Interpret the Data

Copy your final equation here. $y =$ _____
Answer the following questions. Show your work.

1. Rewrite the equation to express the temperature as a function of time elapsed.

    _____ = _____

2. If you were to measure the temperature in degrees Fahrenheit, what effect
    do you think it would have on your curve?

    _____

    _____

    Predict how C and D would change. C_____ D_____

3. Solve your first equation for x.

    $x =$ _____

4. Write the new equation, giving time (the independent variable) as a
    function of the temperature (the dependent variable).

    Time _____

5. Use a graphing calculator to graph both functions on the same set of axes.
    Copy the graphs below. Indicate the x and y ranges.

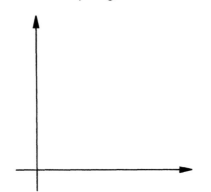

The Rising Damp

In this experiment, the elapsed time is a function of how far the water has risen. The water level is measured in units marked on a strip of blotting paper. The mark number on the strip of blotting paper is the *independent variable,* and the time it takes the water to reach the mark number is the *dependent variable.*

Equipment

strips of blotting paper about $5 \times \frac{3}{4}$ inches, 1 per group

small cups or glasses of water, 1 per group

Coloring the water does not make the waterline more visible; the dye rises more slowly than the water.

clothespins or paper clips, 1 per group

rulers, 1 per group

stopwatches, or watches that display seconds, 1 per group

graph paper, 1 sheet per student

Procedure

Each group marks along the edges of a strip of blotting paper in either centimeters or quarter-inch units (with quarter-inch marks, students will have more data points with which to work). The bottom mark should be about three-quarters of an inch from one end. Students then number the marks, starting from 0 with the bottom mark.

Next, students use clothespins or paper clips to attach the strip of blotting paper so that it can be suspended over the water. They note where the bottom of the strip is, remove the strip, and add water to the glass so that the bottom half inch of the strip will be submerged. They suspend the strip in the water and start the stopwatch when the water reaches mark 0. As the water level rises to each successive mark, they record the time.

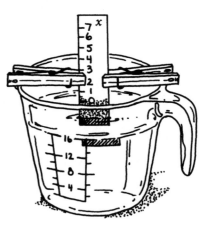

Remind students to convert minutes and seconds to seconds. Time varies with the type of blotting paper. It can take more than 10 minutes for the water to rise 8 centimeters or 3 inches.

Students will notice that it takes the water twice as long to reach the next mark as it did the last mark. This is typical exponential behavior. Students will probably find their graph is of the form $y = Ce^{Dx}$.

Experiment 10

The Rising Damp

Name _____

Partner _____

Collect the Data

Draw a diagram of the experiment, indicating variables

Describe the procedure for the experiment.

The independent variable, x, is _____   Units _____

The dependent variable, y, is _____   Units _____

Equipment (labels and measurements)

Data Collection	
Independent	Dependent

Points to Be Graphed	
x	y

Experiment 10

The Rising Damp

Name_____

Find the Equation

Plot the points you recorded.

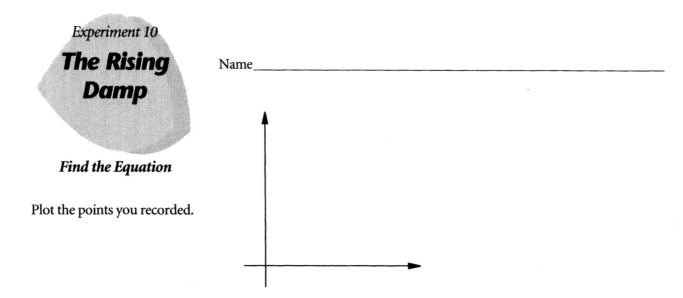

Compare your data to these basic graphs. Circle the graphs to which these points might belong.

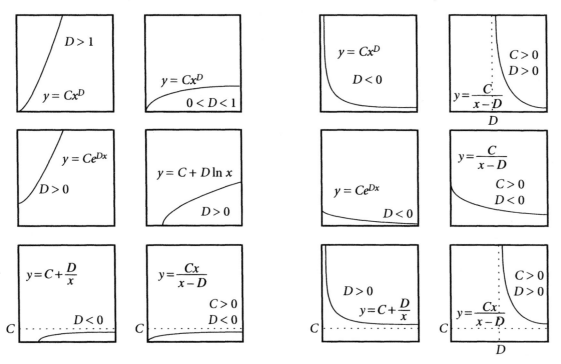

Which basic graph does your data most closely match? $y = $ _____
Explain the reasons for your choice.

_____

_____

Do you have second choice? $y = $ _____

Why or why not? _____

_____

Experiment 10

The Rising Damp

Name _____

Find the Equation, page 2

State the test equation you are using. $y =$ _____

Using either graphing calculator or computer, test values of C and D until you think you are close. List at least four sets of values, and sketch their graphs along with the data points. If you use a computer, state the value of the error.

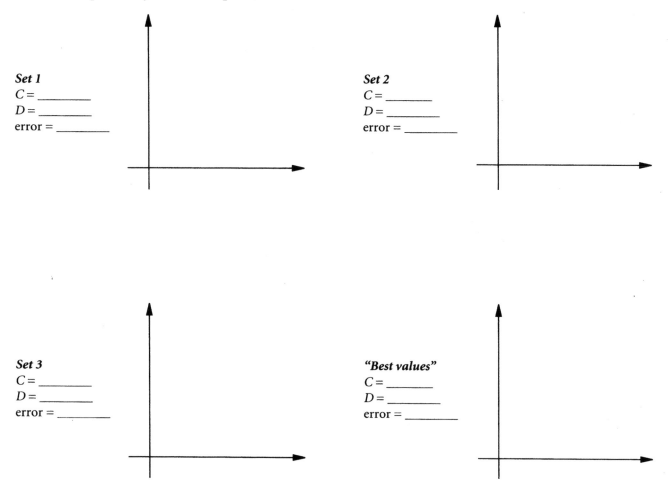

Set 1
$C =$ _____
$D =$ _____
error = _____

Set 2
$C =$ _____
$D =$ _____
error = _____

Set 3
$C =$ _____
$D =$ _____
error = _____

"Best values"
$C =$ _____
$D =$ _____
error = _____

Experiment 10

The Rising Damp

Name _____

Find the Equation, page 3

State the test equation you are investigating. $y =$ _____
Show how your test equation in x and y can be
transformed into a linear equation in T and W.

Use a calculator or computer program to find the best
linear approximation to the transformed points.

Transformed independent variable $T =$ _____

Transformed dependent variable $W =$ _____

List the transformed points, and then graph them.

Independent Variable T	Dependent Variable W

Using algebra, find the equation of the function that
best approximates your *original* data. Show your work.

Experiment 10

The Rising Damp

Name _____

Interpret the Data

Copy your final equation here. $y =$ _____

Answer the following questions. Show your work.

1. Rewrite the equation to express time as a function of the mark number.

    _____ = _____

2. If your marks on the blotter were farther apart, what effect do you think it would have on your curve?

    _____

    _____

    Predict how the equation would change. _____

3. Solve your first equation for x.

    $x =$ _____

4. Write the new equation, giving mark number (the independent variable) as a function of time (the dependent variable).

    Mark number = _____

5. Use a graphing calculator to graph both functions on the same set of axes. Copy the graphs below. Indicate the x and y ranges.

Mirror, Mirror on the Floor

Teaching Notes

In this experiment, the distance from the student to the mirror is a function of the height of the mark that the student is trying to see in the mirror. The height of a mark above the floor is the *independent variable,* and the distance from the mirror a student must stand to see the mark's reflection in the mirror is the *dependent variable.*

Equipment

small, flat mirrors, 1 per group

yardsticks, or poles marked at approximately 6-inch intervals, 2 per group

graph paper, 1 sheet per student

Procedure

Students place a yardstick or marked pole against the wall and then place a mirror 6 to 10 inches from the wall. Standing on the other side of the mirror, they determine the distance from the mirror they must be to see the reflections of marks at given levels above the floor.

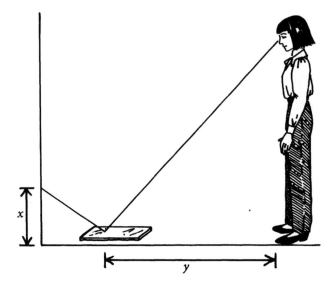

Students must decide whether to measure distance when the mark appears at the edge of the mirror or when the mark appears at its center. Measuring the distance from the toes to the edge or center of the mirror provides sufficient accuracy. The results depend on the mirror and method used, but should be close to the idealized answer of $y = \frac{c}{x}$, which is implied by the diagram:

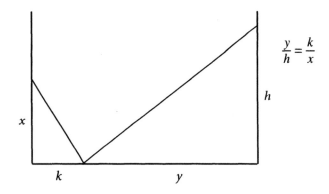

$$\frac{y}{h} = \frac{k}{x}$$

Experiment 11

Mirror, Mirror on the Floor

Name _____

Partner _____

Collect the Data

Draw a diagram of the experiment, indicating variables.

Describe the procedure for the experiment.

_____

_____

_____

_____

_____

_____

_____

_____

_____

_____

_____

_____

The independent variable, x, is _____   Units _____

The dependent variable, y, is _____   Units _____

Equipment (labels and measurements)

Pole or yardstick _____

Mirror _____   Dimensions of mirror _____

Data Collection
Independent	Dependent

Points to Be Graphed
x	y

Mirror, Mirror on the Floor

Name_____

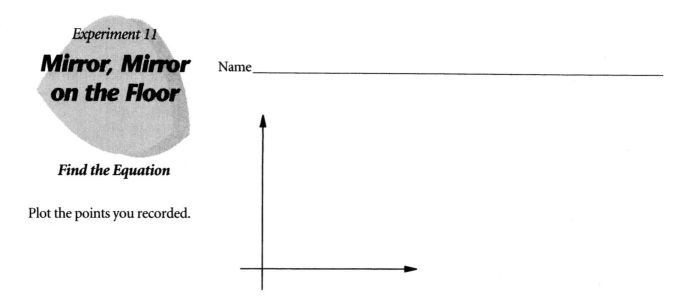

Find the Equation

Plot the points you recorded.

Compare your data to these basic graphs. Circle the graphs to which these points might belong.

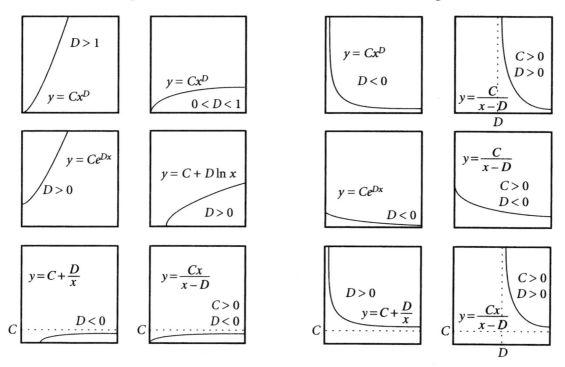

Increasing Functions

$D > 1$

$y = Cx^D$

$y = Cx^D$

$0 < D < 1$

$y = Ce^{Dx}$

$D > 0$

$y = C + D \ln x$

$D > 0$

$y = C + \dfrac{D}{x}$

C · · · · · · · $D < 0$

$y = \dfrac{Cx}{x - D}$

C · · · · · · · $C > 0$ $D < 0$

Decreasing Functions

$y = Cx^D$

$D < 0$

$C > 0$ $D > 0$

$y = \dfrac{C}{x - D}$

D

$y = \dfrac{C}{x - D}$

$C > 0$ $D < 0$

$y = Ce^{Dx}$

$D < 0$

$D > 0$

$y = C + \dfrac{D}{x}$

C · · · · · · · ·

$y = \dfrac{Cx}{x - D}$

$C > 0$ $D > 0$

C · · · · · · · ·

D

Which basic graph does your data most closely match? $y =$ _____

Explain the reasons for your choice.

_____

_____

Do you have second choice? $y =$ _____

Why or why not? _____

_____

84

Experiment 11

Mirror, Mirror on the Floor

Name _____

Find the Equation, page 2

State the test equation you are using. $y =$ _____

Using either graphing calculator or computer, test values of C and D until you think you are close. List at least four sets of values, and sketch their graphs along with the data points. If you use a computer, state the value of the error.

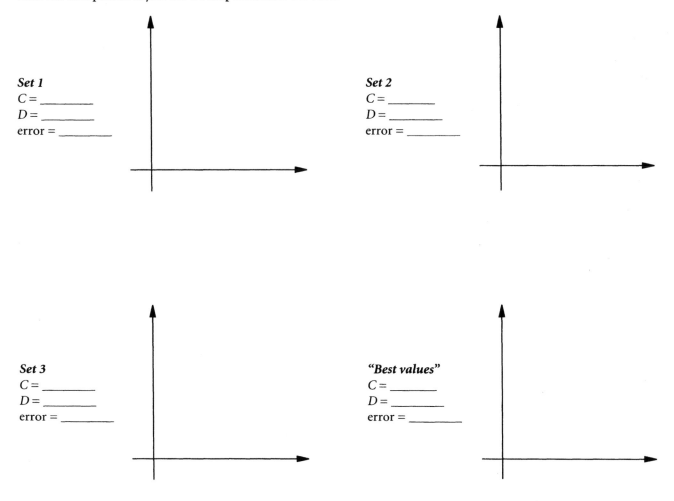

Set 1
$C =$ _____
$D =$ _____
error = _____

Set 2
$C =$ _____
$D =$ _____
error = _____

Set 3
$C =$ _____
$D =$ _____
error = _____

"Best values"
$C =$ _____
$D =$ _____
error = _____

Mirror, Mirror on the Floor

Name _____

Find the Equation, page 3

State the test equation you are investigating. $y =$ _____
Show how your test equation in x and y can be transformed into a linear equation in T and W.

Use a calculator or computer program to find the best linear approximation to the transformed points.

Transformed independent variable $T =$ _____

Transformed dependent variable $W =$ _____

List the transformed points, and then graph them.

Independent Variable T	Dependent Variable W

Using algebra, find the equation of the function that best approximates your *original* data. Show your work.

Experiment 11

Mirror, Mirror on the Floor

Name _____

Interpret the Data

Copy your final equation here. $y =$ _____

Answer the following questions. Show your work.

1. Rewrite the equation to express the distance from the mirror as a function of the height of the mark.

    _____ = _____

2. If y were the distance from the wall to your toes, what effect would that have on your curve?

    _____

    _____

    Predict how C and D would change. C _____ D _____

3. Solve your first equation for x.

    $x =$ _____

4. Write the new equation, expressing the height of the mark (the independent variable) as distance from the mirror (the dependent variable).

    Mark height = _____

5. Use a graphing calculator to graph both functions on the same set of axes. Copy the graphs below. Indicate the x and y ranges.

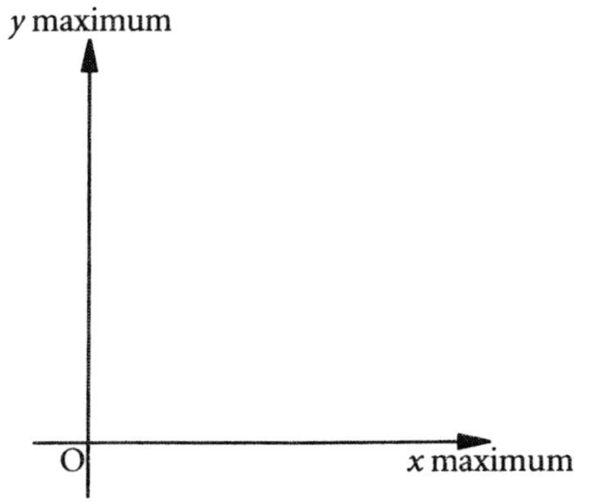

Experiment 12
Musical Glasses

Teaching Notes

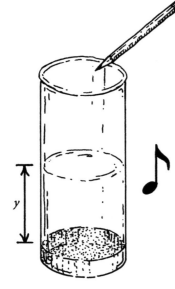

In this experiment, the height of the water is a function of the musical note. The number of a musical note is the *independent variable,* and the water level in a glass is the *dependent variable.*

Equipment

straight-sided drinking glasses, 1 per group

Glasses should be as tall as possible, but at least 7 inches tall.

centimeter rulers, 1 per group

water

electronic tuner, small electronic keyboard, or piano

wooden pencil, 1 per group

graph paper, 1 sheet per student

Procedure

Before conducting this experiment, students will need to understand the difference between whole and half tones on the scale. There should be at least 3 cm of water in the glass for the base tone. On any keyboard (electronic or piano), adjacent keys, black or white, are one half tone apart.

Groups will have to take turns collecting data. Have students add water to the glass and tap it with a pencil until a recognizable note is heard when the glass is struck. Students continue adding water, noting the depth as each whole tone is reached. (Tones become lower as water is added.)

A sample trial yielded the following data:

Note	Note Number	Water Level
F	1	6.7 cm
E flat	2	7.8 cm
D flat	3	9.4 cm
B	4	11.6 cm

If the note number is the independent variable and the water level is the dependent variable, calculated regression gives:

$$y = 5.49^{0.18x}$$

with a regression coefficient of $r = 0.997$.

Experiment 12

Musical Glasses

Name _____

Partner _____

Collect the Data

Draw a diagram of the experiment, indicating variables.

Describe the procedure for the experiment.

_____

_____

_____

_____

_____

_____

_____

_____

_____

_____

The independent variable, x, is _____   Units _____

The dependent variable, y, is _____   Units _____

Equipment (labels and measurements)

Glass _____   Height _____

Diameter _____

Data Collection			Points to Be Graphed	
Independent	Dependent		x	y

Experiment 12
Musical Glasses

Name_____

Find the Equation

Plot the points you recorded.

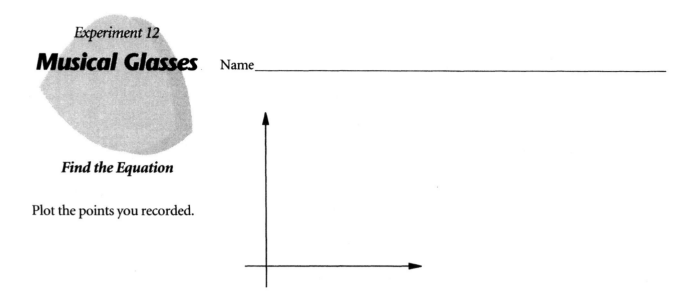

Compare your data to these basic graphs. Circle the graphs to which these points might belong.

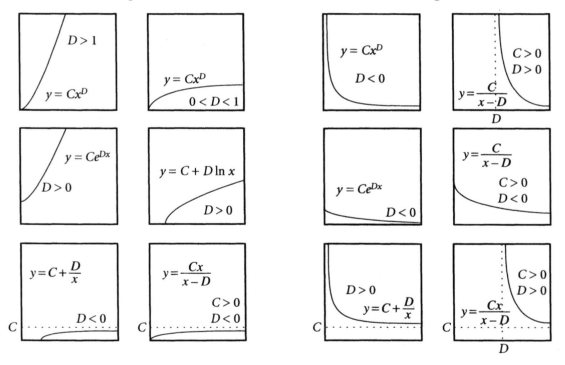

Increasing Functions

$D > 1$
$y = Cx^D$

$y = Cx^D$
$0 < D < 1$

$y = Ce^{Dx}$
$D > 0$

$y = C + D \ln x$
$D > 0$

$y = C + \dfrac{D}{x}$
$D < 0$
C

$y = \dfrac{Cx}{x - D}$
$C > 0$
$D < 0$
C

Decreasing Functions

$y = Cx^D$
$D < 0$

$y = \dfrac{C}{x - D}$
$C > 0$
$D > 0$
D

$y = \dfrac{C}{x - D}$
$C > 0$
$D < 0$

$y = Ce^{Dx}$
$D < 0$

$D > 0$
$y = C + \dfrac{D}{x}$
C

$y = \dfrac{Cx}{x - D}$
$C > 0$
$D > 0$
C
D

Which basic graph does your data most closely match? $y =$ _____
Explain the reasons for your choice.

_____

_____

Do you have second choice? $y =$ _____

Why or why not? _____

_____

Experiment 12

Musical Glasses   Name _____

Find the Equation, page 2

State the test equation you are using. $y =$ _____

Using either graphing calculator or computer, test values of C and D until you
think you are close. List at least four sets of values, and sketch their graphs along
with the data points. If you use a computer, state the value of the error.

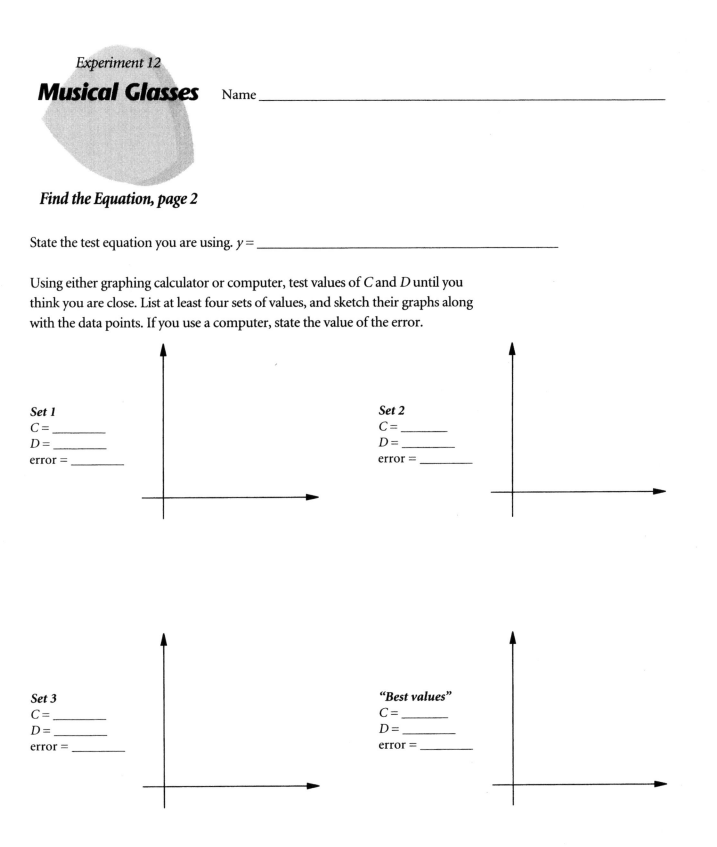

Set 1
$C =$ _____
$D =$ _____
error = _____

Set 2
$C =$ _____
$D =$ _____
error = _____

Set 3
$C =$ _____
$D =$ _____
error = _____

"Best values"
$C =$ _____
$D =$ _____
error = _____

Experiment 12

Musical Glasses

Name _____

Find the Equation, page 3

State the test equation you are investigating. $y =$ _____
Show how your test equation in x and y can be
transformed into a linear equation in T and W.

Use a calculator or computer program to find the best
linear approximation to the transformed points.

Transformed independent variable $T =$ _____

Transformed dependent variable $W =$ _____

List the transformed points, and then graph them.

Independent Variable T	Dependent Variable W

Using algebra, find the equation of the function that
best approximates your *original* data. Show your work.

Experiment 12

Musical Glasses

Name _____

Interpret the Data

Copy your final equation here. $y =$ _____

Answer the following questions. Show your work.

1. Rewrite the equation to express the water level as a function of the note.

   _____ = _____

2. If you were to use taller glass, what effect would that have on your curve?

   _____

   _____

   Predict how C and D would change. C_____ D_____

3. Solve your first equation for x.

   $x =$ _____

4. Write the new equation, expressing the note number (the independent variable) as a function of water level (the dependent variable).

   Note number _____

5. Use a graphing calculator to graph both functions on the same set of axes. Copy the graphs below. Indicate the x and y ranges.

Experiment 13

Balancing Act

Teaching Notes

In this experiment, the number of weights is a function of the distance from a pivot hole to a bucket. The distance from the pivot hole to the bucket is the *independent variable,* and the number of weights needed to make the ruler balance is the *dependent variable.*

Equipment

rulers with center hole, 1 per group

film containers, 1 per group

anchors

> *Use heavier weights, such as nuts or washers, suspended from a paper clip. The anchor should be of a different material than the weights used to measure.*

light weights

> *The weights should be light enough so that near-balance can be achieved from almost any position. Use candy-coated chocolates, X-shaped paper clips, or any other small objects.*

large paper clips or string

> *Used to attach the film container and the anchors to the ruler.*

pencil, 1 per group

graph paper, 1 sheet per student

Procedure

Each group decides on the location and weight of the "anchor." These are fixed constants. Students should firmly attach the anchor to the ruler and then select the first location on the other side of the pivot hole to hang the bucket (film container) of weights. The bucket of weights is attached by threading either string or large paper clips through two holes punched opposite one another near the rim of the film container. The assembly must hang freely. Students now make a balance by inserting the pencil into one of the round holes in the ruler.

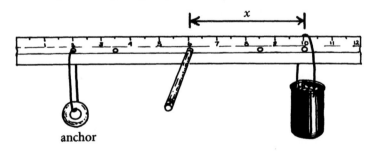

anchor

In theory, $yx = $ constant, or $y = \dfrac{C}{x}$. The experimental factors are likely to lead to $y = Cx^D$, with $-2 < D < 0$.

Experiment 13

Balancing Act

Name _____

Partner _____

Collect the Data

Draw a diagram of the experiment, indicating variables.

Describe the procedure for the experiment.

_____

_____

_____

_____

_____

_____

_____

_____

_____

_____

_____

The independent variable, x, is _____ Units _____

The dependent variable, y, is _____ Units _____

Equipment (labels and measurements)

Anchor position _____

Type of weight _____ Weight numbers _____

<table>
<tr><th colspan="2">*Data Collection*</th><th colspan="2">*Points to Be Graphed*</th></tr>
<tr><th>Independent</th><th>Dependent</th><th>x</th><th>y</th></tr>
<tr><td></td><td></td><td></td><td></td></tr>
<tr><td></td><td></td><td></td><td></td></tr>
<tr><td></td><td></td><td></td><td></td></tr>
<tr><td></td><td></td><td></td><td></td></tr>
<tr><td></td><td></td><td></td><td></td></tr>
<tr><td></td><td></td><td></td><td></td></tr>
</table>

Balancing Act

Name_____

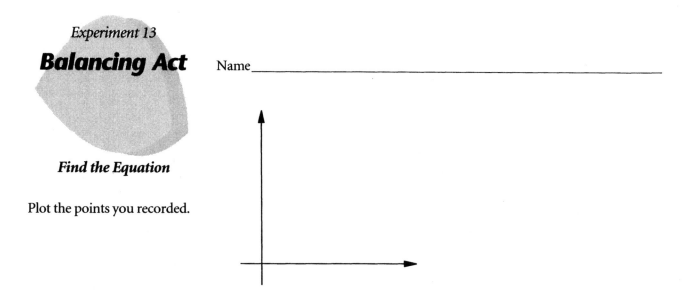

Find the Equation

Plot the points you recorded.

Compare your data to these basic graphs. Circle the graphs to which these points might belong.

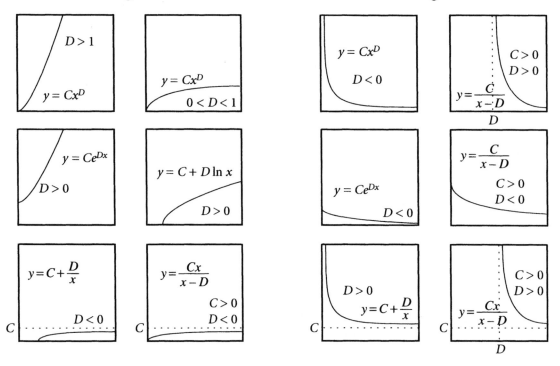

Increasing Functions	*Decreasing Functions*

$y = Cx^D$, $D > 1$

$y = Cx^D$, $0 < D < 1$

$y = Ce^{Dx}$, $D > 0$

$y = C + D \ln x$, $D > 0$

$y = C + \dfrac{D}{x}$, $D < 0$

$y = \dfrac{Cx}{x - D}$, $C > 0$, $D < 0$

$y = Cx^D$, $D < 0$

$y = \dfrac{C}{x - D}$, $C > 0$, $D > 0$

$y = \dfrac{C}{x - D}$, $C > 0$, $D < 0$

$y = Ce^{Dx}$, $D < 0$

$y = C + \dfrac{D}{x}$, $D > 0$

$y = \dfrac{Cx}{x - D}$, $C > 0$, $D > 0$

Which basic graph does your data most closely match? $y =$ _____
Explain the reasons for your choice.

_____

_____

Do you have second choice? $y =$ _____
Why or why not? _____

_____

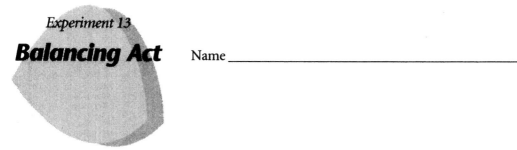

Experiment 13

Balancing Act

Name _____

Find the Equation, page 2

State the test equation you are using. $y =$ _____

Using either graphing calculator or computer, test values of C and D until you
think you are close. List at least four sets of values, and sketch their graphs along
with the data points. If you use a computer, state the value of the error.

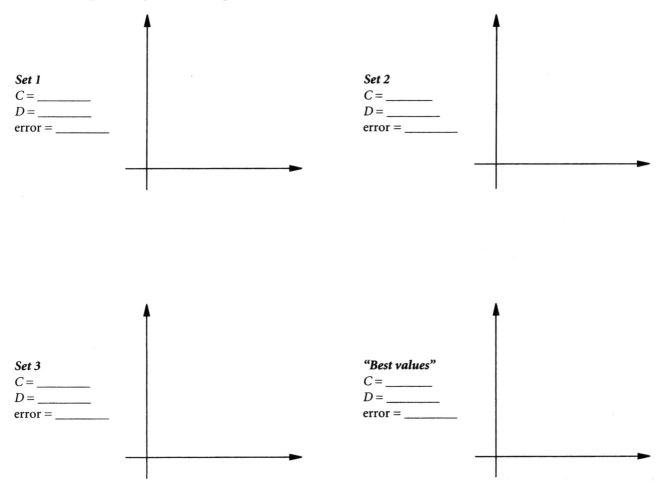

Set 1
$C =$ _____
$D =$ _____
error = _____

Set 2
$C =$ _____
$D =$ _____
error = _____

Set 3
$C =$ _____
$D =$ _____
error = _____

"Best values"
$C =$ _____
$D =$ _____
error = _____

Experiment 13

Balancing Act

Name _____

Find the Equation, page 3

State the test equation you are investigating. $y =$ _____

Show how your test equation in x and y can be transformed into a linear equation in T and W.

Use a calculator or computer program to find the best linear approximation to the transformed points.

Transformed independent variable $T =$ _____

Transformed dependent variable $W =$ _____

List the transformed points, and then graph them.

Independent Variable T	Dependent Variable W

Using algebra, find the equation of the function that best approximates your *original* data. Show your work.

98

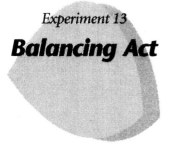

Experiment 13

Balancing Act

Name _____

Interpret the Data

Copy your final equation here. $y =$ _____

Answer the following questions. Show your work.

1. Rewrite the equation to express the number of weights as a function of the distance from the center.

    _____ = _____

2. If you were to increase the anchor weight, what effect would that have on your curve?

    _____

    _____

    _____

    Predict how *C* and *D* would change. C_____ D _____

3. Solve your first equation for *x*.

    $x =$ _____

4. Write the new equation, giving distance from the center (the independent variable) as a function of the number of weights (the dependent variable).

    Distance = _____

5. Use a graphing calculator to graph both functions on the same set of axes. Copy the graphs below. Indicate the *x* and *y* ranges.

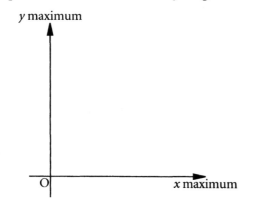

y maximum

O

x maximum

Experiment 14
Falling Marbles

Teaching Notes

In this experiment, the distance that the marble travels from the base of the table is a function of the height of the ramp. The height of the ramp is the *independent variable*, and the distance at which a marble lands from the base of the table is the *dependent variable*.

Equipment

marbles, 1 per group

yardsticks, 1 per group

ramps, 1 per group

> *Number the ramps. If they are made from slats of wood, bevel the downhill end. Vinyl gutters are inexpensive; most building-supply stores will cut up a 10-foot length for you.*

blocks, books, or another material to raise the ramps

graph paper, 1 sheet per student

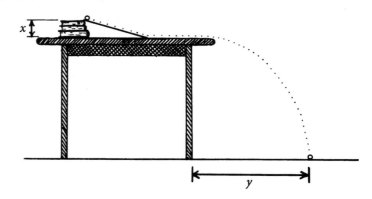

Procedure

Have each group set up a ramp on a table so the lower edge of the ramp is about 1 foot from the edge of the table. Groups then choose a height for the ramp (the independent variable) and allow the marble to roll down the ramp. They must start the marble at the top of the ramp for each trial. They record the distance from the base of the table to where the marble first hits (the dependent variable). Groups repeat the procedure with differing ramp heights.

Falling Marbles

Name _____

Partner _____

Collect the Data

Draw a diagram of the experiment, indicating variables.

Describe the procedure for the experiment.

_____

_____

_____

_____

_____

_____

_____

_____

_____

_____

The independent variable, *x*, is _____   Units _____

The dependent variable, *y*, is _____   Units _____

Equipment (labels and measurements)

Ramp _____   Ramp length _____

Table height_____   Marble diameter _____

Data Collection

Independent	Dependent

Points to Be Graphed

x	*y*

Experiment 14
Falling Marbles

Name_____

Find the Equation

Plot the points you recorded.

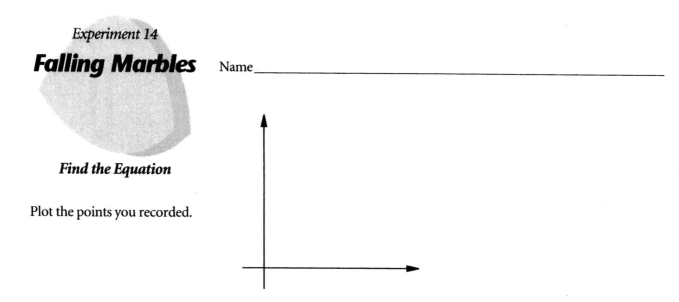

Compare your data to these basic graphs. Circle the graphs to which these points might belong.

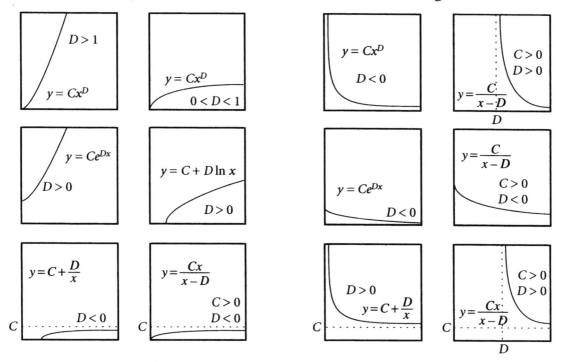

Which basic graph does your data most closely match? $y =$ _____
Explain the reasons for your choice.

_____

_____

Do you have second choice? $y =$ _____

Why or why not? _____

_____

Experiment 14

Falling Marbles

Name _____

Find the Equation, page 2

State the test equation you are using. $y = $ _____

Using either graphing calculator or computer, test values of C and D until you
think you are close. List at least four sets of values, and sketch their graphs along
with the data points. If you use a computer, state the value of the error.

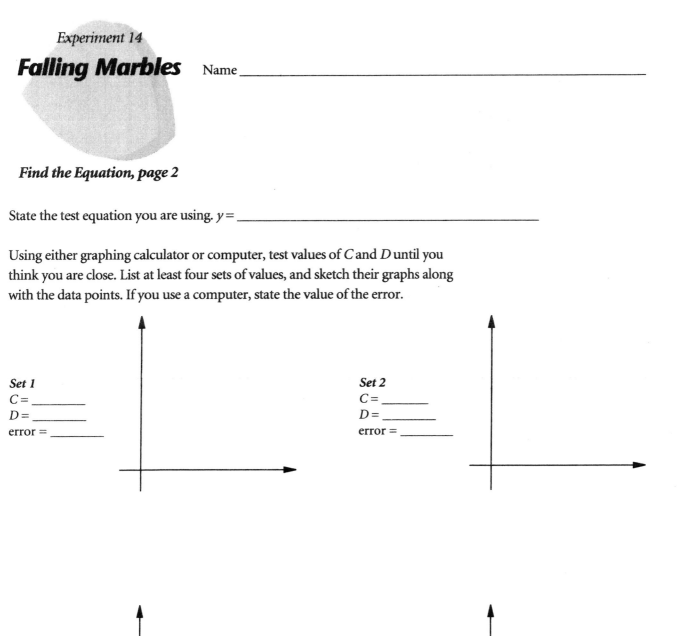

Set 1
$C = $ _____
$D = $ _____
error = _____

Set 2
$C = $ _____
$D = $ _____
error = _____

Set 3
$C = $ _____
$D = $ _____
error = _____

"Best values"
$C = $ _____
$D = $ _____
error = _____

Experiment 14

Falling Marbles

Name _____

Find the Equation, page 3

State the test equation you are investigating. $y =$ _____

Show how your test equation in x and y can be transformed into a linear equation in T and W.

Use a calculator or computer program to find the best linear approximation to the transformed points.

Transformed independent variable $T =$ _____

Transformed dependent variable $W =$ _____

List the transformed points, and then graph them.

Independent Variable T	Dependent Variable W

Using algebra, find the equation of the function that best approximates your *original* data. Show your work.

104

Experiment 14

Falling Marbles

Name _____

Interpret the Data

Copy your final equation here. $y =$ _____

Answer the following questions. Show your work.

1. Rewrite the equation to distance from the table as a function of the height of the ramp.

   _____ = _____

2. If the height of your table were reduced, what effect would that have on your curve?

   _____

   _____

   Predict how C and D would change. C_____ D_____

3. Solve your first equation for x.

   $x =$ _____

4. Write the new equation, expressing the ramp height (the independent variable) as a function of the distance from the table (the dependent variable).

   Ramp height = _____

5. Use a graphing calculator to graph both functions on the same set of axes. Copy the graphs below. Indicate the x and y ranges.

Appendix A

Applesoft BASIC Listing

```
10 DIM X(10),Y(10)
20 HOME
30 INPUT "HOW MANY POINTS ";H
40 IF H > 10 OR H < 2 THEN 30
50 FOR A = 1 TO H
60 PRINT "INPUT POINT # ";A;" ";
70 INPUT "";X(A),Y(A)
80 NEXT A
90 HOME
100 FOR A = 1 TO H
110 PRINT "POINT #";A,X(A);",";Y(A)
120 NEXT A
130 INPUT "ARE THESE POINTS CORRECT?";A$
140 IF A$ = "Y" OR A$ = "y" OR A$ = "YES"
    THEN 180
150 INPUT "WHICH POINT NEEDS TO BE
    CORRECTED?";W
160 INPUT "TYPE IN THE NEW COORDINATES
    ";X(W),Y(W)
170 GOTO 90
180 INPUT "PLEASE ENTER THE MAXIMUM VALUE FOR
    X";XR
190 INPUT "PLEASE ENTER THE MAXIMUM VALUE FOR
    Y";YR
200 SX = 261 / (XR + 1)
210 SY = 160 / (YR + 1)
220 HGR
230 VTAB 21
240 PRINT "1) PLOT POINTS ON A CLEAN SCREEN"
250 PRINT "2) INPUT VALUES FOR C AND D"
260 PRINT "3) QUIT";
270 INPUT Q
280 ON Q GOSUB 300,390,560
290 HOME : GOTO 230
```

```
300 HGR
310 HCOLOR= 3
320 HPLOT 10,0 TO 10,159 TO 271,159
330 FOR A = 1 TO H
340 X = INT (X(A) * SX + 10 + .5)
350 Y = INT (159 - SY * Y(A) + .5)
360 HPLOT X,Y TO X - 1,Y TO X - 1,Y - 1 TO
    X,Y - 1 TO X,Y
370 NEXT A
380 RETURN
390 INPUT "TYPE IN VALUES FOR C & D IN
    Y=Cx^D";C,D
    (replace this line with other functions)
400 HPLOT 10,159
    410 FOR Q = 11 TO 271 STEP 3
420 X = (Q - 10) / SX
430 Y = C * X ^D
    (replace this line with other functions)
440 T = 159 - SY * Y
450 IF T < 0 OR T > 159 THEN 470
460 HPLOT TO Q,T
470 NEXT Q
480 S=0
490 FOR I = 1 TO H
500 W = C * X(I) ^ D: REM to change function
510 S=S+ (W-Y(I)) ^ 2
520 NEXT I
530 E= SQR(S)
540 PRINT "C=";C;" D= ";D;" ERROR = ";E:
    PRINT "PRESS RETURN TO CONTINUE";: GET
    T$
550 RETURN
560 TEXT
570 END
```

Appendix B

Macintosh QuickBASIC™ Listing

```
REM Ron Carlson
REM Canton H.S. Canton, Mich 48187

GOTO 10
1 DEF FNf(z)=c*z^d
fun$="Y= CX^D"
RETURN
2 DEF FNf(z) = c*EXP(d*z)
fun$="Y= Ce^DX"
RETURN
3 DEF FNf(z)= c+d*LOG(z)
fun$="Y= C+D lnX"
RETURN
4 DEF FNf(z)= c+d/z
fun$="Y= C+D/X"
RETURN
5 DEF FNf(z)= c/(z-d)
fun$="Y= C/(X-D)"
RETURN
6 DEF FNf(z)= c*z/(z-d)
fun$="Y= CX/(X-D)"
RETURN

10 DIM x(15),y(15),incr(7)
 incr(1)=1000
 incr(2)=100
 incr(3)=10
 incr(4)=1
 incr(5)=.1
 incr(6)=.01
 incr(7)=.001
20 WINDOW 1,,(1,20)-(500,250),3

PRINT"FUNCTION TESTER-ALGEBRA EXPERIMENTS"
PRINT

PRINT "FIRST, ENTER THE NUMBER OF THE TEST
    FUNCTION (return)"
PRINT "THEN ENTER THE DATA POINTS"
PRINT
PRINT "Clean up a cluttered screen with PLOT
    PTS"
PRINT

PRINT "1  Y=CX^D"
PRINT "2  Y=Ce^DX"
PRINT"3  Y=C+D lnX"
PRINT"4  Y=C+D/X"
PRINT"5  Y=C/(X-D)"
```

```
PRINT"6  Y=CX/(X-D)"
21 INPUT"which function ";f
22 ON f GOSUB 1,2,3,4,5,6
30 INPUT "HOW MANY DATA POINTS ";h
40 IF h > 15 OR h < 2 THEN 30
50 FOR a = 1 TO h

45 PRINT "ENTER THE POINTS - SEPARATE THE
    COORDINATES WITH A COMMA"

60 PRINT "INPUT POINT # ";a;" ";
70 INPUT "";x(a),y(a)
80 NEXT a
90 CLS
100 FOR a = 1 TO h
110 PRINT "POINT #";a,x(a);",";y(a)
120 NEXT a
130 INPUT "ARE THESE POINTS CORRECT?";a$
140 IF a$ = "Y" OR a$ = "y" OR a$ = "YES"
    THEN 180
150 INPUT "WHICH POINT NEEDS TO BE
    CORRECTED?";w
160 INPUT "TYPE IN THE NEW COORDINATES
    ";x(w),y(w)
170 GOTO 90
180 INPUT "PLEASE ENTER THE MAXIMUM VALUE FOR
    X";XR
190 INPUT "PLEASE ENTER THE MAXIMUM VALUE FOR
    Y";YR
195 CLS
200 SX = 261 / (XR + 1)
210 SY = 200 / (YR + 1)

WINDOW 2,,(20,260)-(500,330),2
PRINT "C "
LOCATE 1,18
PRINT "Incr"
LOCATE 1,43
PRINT "ERROR"
PRINT "D "
LOCATE 2,18
PRINT"Incr"
LOCATE 2,43
PRINT fun$

BUTTON 1,1,"PLOT PTS.",(2,47)-(75,62),1
BUTTON 2,1,"GRAPH FUNCTION",(85,47)-(200,62)
BUTTON 3,1,"QUIT",(250,47)-(340,62),1
BUTTON 4,1,"+",(20,0)-(30,8),1
BUTTON 5,1,"-",(20,9)-(30,17),1
```

```
BUTTON 6,1,"+",(165,0)-(176,8),1            CASE ELSE
BUTTON 7,1,"-",(165,9)-(176,17),1           END SELECT
BUTTON 8,1,"+",(20,18)-(30,26),1            WEND
BUTTON 9,1,"-",(20,27)-(30,35),1            but%=DIALOG (1)
BUTTON 10,1,"+",(165,18)-(176,26),1          IF but%=1 THEN GOSUB 300
BUTTON 11,1,"-",(165,27)-(176,35),1          IF but%=2 THEN GOSUB 385
                                             IF but% <> 3 THEN WINDOW 2
but%=0                                      LOCATE 1,50
NOTFINISHED%=1                               PRINT USING tmp$;SQR(s);
EVENT%=0                                    NOTFINISHED%=1
c=1                                        WEND
d=1
cincr=4                                    GOTO 490
dincr=4
WHILE but% <> 3                            300 WINDOW 1
 WHILE NOTFINISHED%                        CLS
 EVENT%=DIALOG(0)                          320 LINE( 10,0)- (10,200),1,bf
 SELECT CASE EVENT%                        325 LINE (10,200)- ( 271,200),1,bf
 CASE 1                                     330 FOR a = 1 TO h
  NOTFINISHED%=0                            340 x = INT (x(a) * SX + 10 + .5)
IF DIALOG(1)=6 THEN cincr=cincr-1          350 y = INT (200 - SY * y(a) + .5)
IF DIALOG(1)=7 THEN cincr=cincr+1          360 CIRCLE (x,y),2
IF cincr<1 THEN cincr=1                     370 NEXT a
IF cincr>7 THEN cincr=7                     380 RETURN
                                           385 WINDOW 1
tmp$="#####.###"                           410 FOR q = 11 TO 271 STEP 3
LOCATE 1,25                                420 x = (q - 10) / SX
PRINT  incr(cincr)                         IF f>=5 AND x=d THEN 470
                                           430 y = FNf(x)
IF DIALOG(1)=10 THEN dincr=dincr-1         440 dd = 200 - SY * y
IF DIALOG(1)=11 THEN dincr=dincr+1         450 IF dd < 0 OR dd > 200 THEN 470
IF dincr<1 THEN dincr=1                    460 PSET( q,dd),1
IF dincr>7 THEN dincr=7                    470 NEXT q
LOCATE 2,25                                477 s=0:w=0
PRINT  incr(dincr)                         479 FOR i =1 TO h
IF DIALOG(1)=4 THEN c=c+incr(cincr)        IF f=4 AND x(i)=0 THEN 484
IF DIALOG(1)=5 THEN c=c-incr(cincr)        IF f>=5 AND x(i)=d THEN 484
LOCATE 1,6                                 480 w=FNf(x(i)):
PRINT USING tmp$;c                         482 s=s+(w-y(i))^2
IF DIALOG(1)=8 THEN d=d+incr(dincr)        484 NEXT i
IF DIALOG(1)=9 THEN d=d-incr(dincr)        488 RETURN
LOCATE 2,6                                 490 END
PRINT USING tmp$;d
```

Sample Output

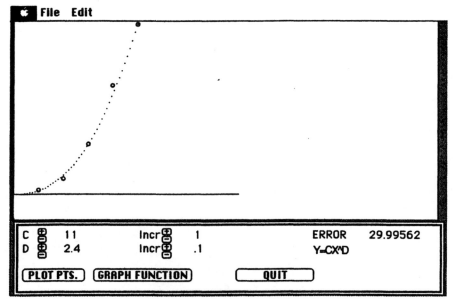